ABOUT THE AUTHOR

Catherine V. Jeremko is a certified secondary mathematics teacher in New York State. She is the author of *Just in Time Math*, has contributed to *501 Quantitative Comparison Questions*, and has edited two publications, *GMAT Success!* and the second edition of *501 Algebra Questions*, all published by LearningExpress. She currently teaches seventh grade mathematics at Vestal Middle School in Vestal, New York. Ms. Jeremko is also a teacher trainer for the use of technology in the mathematics classroom. She resides in Apalachin, New York with her three daughters.

CONTENTS

FORMULA CHEAT SHEET

PERIMETER
Rectangle: $P = 2 \times l + 2 \times w$
Square: $P = 4 \times s$

CIRCUMFERENCE OF A CIRCLE
$C = \pi \times d$ or $C = 2 \times \pi \times r$

AREA
Triangle: $A = \frac{1}{2} \times b \times h$ **Rectangle:** $A = b \times h$
Trapezoid: $A = \frac{1}{2} \times h \times (b_1 + b_2)$ **Circle:** $A = \pi r^2$

SURFACE AREA
Rectangular Prism: $SA = 2(l \times w) + 2(l \times h) + 2(w \times h)$
Cube: $SA = 6s^2$
Cylinder: $2(\pi r^2) + 2\pi rh$
Sphere: $4\pi r^2$

VOLUME $V = b \times h$ (the area of the base times the height)
Rectangular Solid: $V = l \times w \times h$
Cylinder: $V = \pi \times r^2 \times h$
Cube: $V = s^3$
Triangular Prism: $\frac{1}{2} bh_1 \times h_2$
Trapezoidal Prism: $\frac{1}{2} h_1 (b_1 + b_2) \times h_2$
Sphere: $\frac{4}{3} \pi r^3$
Pyramid: $\frac{1}{3} lwh$
Cone: $\frac{1}{3} \pi r^2 h$

PYTHAGOREAN THEOREM
$\text{leg}^2 + \text{leg}^2 = \text{hypotenuse}^2$, or $a^2 + b^2 = c^2$

TRIGONOMETRIC RATIOS
(SOH – CAH – TOA)

Sine of an angle: $\dfrac{\text{length of opposite side}}{\text{length of hypotenuse}}$

Cosine of an angle: $\dfrac{\text{length of adjacent side}}{\text{length of hypotenuse}}$

Tangent of an angle: $\dfrac{\text{length of opposite side}}{\text{length of adjacent side}}$

COORDINATE GEOMETRY

Midpoint of a segment: $M = (\frac{x_1 + x_2}{2}, \frac{y_1 + y_2}{2})$

Distance between two points: $D = \sqrt{(x_2 - x_1)^2 + (y_2 - y_1)^2}$

Slope of a line: $\dfrac{\text{the change in the } y\text{-coordinate's value}}{\text{the change in the } x\text{-coordinate's value}}$, or $\dfrac{\Delta y}{\Delta x}$,

or $\dfrac{y_2 - y_1}{x_2 - x_1}$, or $\dfrac{\text{rise}}{\text{run}}$

Slope-intercept form of a line: $y = mx + b$, where m is the slope

and b is the y-intercept

Introduction

You have to face a big exam that will test your geometry skills. It is just a few weeks, perhaps even just a few days, from now. You haven't begun to study. Perhaps you just haven't had the time. We are all faced with full schedules and many demands on our time, including work, family, and other obligations. Or perhaps you have had the time, but procrastinated; topics in geometry are topics that you would rather avoid at all costs. Formulas and geometric figures have never been your strong suit. It is possible that you have waited until the last minute because you feel rather confident in your mathematical skills, and just want a quick refresher on the major topics. Maybe you just realized that your test included a mathematics section, and now you have only a short time to prepare.

If any of these scenarios sounds familiar, then *Just in Time Geometry* is the right book for you. Designed specifically for last-minute test preparation, *Just in Time Geometry* is a fast, accurate way to build the essential skills necessary to tackle formulas and geometry-related problems. This book includes nine chapters of geometry topics, with an additional chapter on study skills to make your time effective. In just ten short chapters, you will get the essentials—just in time for passing your big test.

THE *JUST IN TIME* TEST-PREP APPROACH

At LearningExpress, we know the importance that is placed on test scores. Whether you are preparing for the PSAT, SAT, GRE, GMAT, a civil service exam, or you simply need to improve your fundamental mathematical skills, our *Just in Time* streamlined approach can work for you. Each chapter includes:

- a ten-question benchmark quiz to help you assess your knowledge of the topics and skills in the chapter
- a lesson covering the essential content for the topic of the chapter

- sample problems with full explanations
- calculator tips to make the most of technology on your exam
- specific tips and strategies to prepare for the exam
- a 25-question practice quiz followed by detailed answers and explanations to help you measure your progress

Our *Just in Time* series also includes the following features:

- *Extra Help* sidebars that refer you to other Learning-Express skill builders or other resources, such as Internet sites, that can help you learn more about a particular topic

- *Calculator Tips:* offers hints on how your calculator can help you.

- *Glossary* sidebars with key definitions

- *Rule Book* sidebars highlighting the rules that you absolutely need to know

- *Shortcut* sidebars with tips for reducing your study time—without sacrificing accuracy

- A *Formula Cheat Sheet* with common formulas for last-minute test preparation

Of course, no book can cover every type of problem you may face on a given test. But this book is not just about recognizing specific problem types; it is also about building those essential skills, confidence, and processes that will ensure success when faced with a geometry problem. The topics in this book have been carefully chosen to reflect not only what you are likely to see on an exam, but also what you are likely to come across regularly in books, newspapers, lectures, and other daily activities.

HOW TO USE THIS BOOK

While each chapter can stand on its own as an effective review of mathematical content, this book will be most effective if you complete each chapter in order, beginning with Chapter 1. Chapters 2 and 3 review the basic knowledge of simple geometric figures. Chapters 4 and 5 review common geometric figures. The material in Chapters 6 and 7 is concerned with geometric measurement. Chapters 8, 9, and 10 cover applications of geometry and coordinate geometry. The chapters are arranged such that material covered earlier chapters may be referenced in a later chapter.

Below is a brief outline of each chapter:

- **Chapter 1: Study Skills** reviews fundamental study strategies including budgeting your time, creating a study plan, and using study aids such as flashcards.
- **Chapter 2: Building Blocks of Geometry: Points, Lines, and Angles** reviews the simplest geometric constructs.
- **Chapter 3: Special Angle Pairs and Angle Measurement** reviews basic information about angles and angle relationships.
- **Chapter 4: Triangles** reviews the most common geometric figure and its properties.
- **Chapter 5: Quadrilaterals and Circles** reviews the four-sided polygon and the round geometric figure, and their properties.
- **Chapter 6: Perimeter and Area** reviews both the measurements around a geometric figure and the measurements that cover a figure.
- **Chapter 7: Surface Area and Volume** is concerned with wrapping a three-dimensional solid and filling a geometric solid.
- **Chapter 8: Transformations and Similarity** reviews the movement of congruent figures and figures that have the same shape, but a different size.
- **Chapter 9: Pythagorean Theorem and Trigonometry** is concerned with two very important applications of the triangle.
- **Chapter 10: Coordinate Geometry** reviews the important relationship between geometry and the coordinate plane.

Depending upon how much time you have before the exam, review as much as possible. If time is short, start your review with the chapters that address your weaknesses. The ten-question benchmark quiz at the start of each chapter can help you assess your strengths and weaknesses.

Finally, remain calm and think positively. Your big test may be just a short while away, but you are taking the steps you need to prepare . . . *just in time*.

Study Skills

If you have left studying for that big test until the last minute, you may be feeling that your only option is to cram. You might be feeling panicky that you will never have enough time to learn what you need to know. But the "Just in Time" solution is exactly that: just in time. This means that with the help of this book you can use your available time prior to your test effectively. First, to get ready for your test just in time, you need a plan. This chapter will help you put together a study plan that maximizes your time and tailors your learning strategy to your needs and goals.

There are four main factors that you need to consider when creating your study plan: what to study, where to study, when to study, and how to study. When you put these four factors together, you can create a specific plan that will allow you to accomplish more—in less time. If you have three weeks, two weeks, or even one week to get ready, you can create a plan that avoids anxiety-inducing cramming and focuses on real learning by following the simple steps in this chapter.

WHAT TO STUDY

Finding out what you need to study for your test is the first step in creating an effective study plan. You need to have a good measure of your ability in geometry. You can accomplish this by looking over the Contents to see what looks familiar to you and by answering the benchmark quiz questions starting in the next chapter. You also need to know what exactly is covered on the test you will be taking. Considering both your ability and the test content will tell you what you need to study.

▶ Establish a Benchmark

In each chapter you will take a short, ten-question benchmark quiz that will help you assess your skills. This may be one of the most important steps in creating your study plan. Because you have limited time, you need to be very efficient in your studies. Once you take a benchmark quiz and analyze the results, you will be able to avoid studying the material you already know. This will allow you to focus on those areas that need the most attention.

A benchmark quiz is only practice. If you do not do as well as you anticipate, do not be alarmed and certainly do not despair. The purpose of the quiz is to help you focus your efforts so that you can *improve*. It is important to analyze your results carefully. Look beyond your score, and consider *why* you answered some questions incorrectly. Here are some questions to ask yourself when you review your wrong answers:

- Did you get the question wrong because the material was totally unfamiliar?
- Was the material familiar but you were unable to come up with the right answer? In this case, when you read the right answer it will often make perfect sense. You might even think, "I knew that!"
- Did you answer incorrectly because you read the question carelessly?
- Did you make another careless mistake? For example, did you circle choice **a** when you meant to circle choice **b**?

Next, look at the questions you answered correctly and review how you came up with the right answer. Not all right answers are created equally.

- Did you simply know the right answer?
- Did you make an educated guess? An educated guess might indicate that you have some familiarity with the subject, but you probably need at least a quick review.
- Did you make a lucky guess? A lucky guess means that you don't know the material and you will need to learn it.

Your performance on the benchmark quiz will tell you several important things. First, it will tell you how much you need to study. For example, if you got eight out of ten questions right (not counting lucky guesses!), you might only need to brush up on certain geometry topics. But if you got five out of ten questions wrong, you will need a thorough review. Second, it can tell you what you know well, that is, which subjects you *don't* need to study. Third, you will determine which subjects you need to study in-depth and which subjects you simply need to review briefly.

▶ Target Your Test

For the "Just in Time" test-taker, it is important to focus your study efforts to match what is needed for your test. You don't want to waste your time learning something that will not be covered on your test. There are three important aspects that you should know about your test before developing your study plan:

- What material is covered?
- What is the format of the test? Is it multiple choice? Fill in the blank? Some combination? Or something else?
- What is the level of difficulty?

How can you learn about the test before you take it? For most standardized tests, there are sample tests available. These tests—which have been created to match the test that you will take—are probably the best way to learn what will be covered. If your test is non-standardized, you should ask your instructor specific questions about the upcoming test.

You should also know how your score will affect your goal. For example, if you are taking the SAT exam, and the median math score of students accepted at your college of choice is 550, then you should set your sights on achieving a score of 550 or better. Or, if you are taking the New York City Police Officer exam, you know that you need to get a perfect or near-perfect score to get a top slot on the list. Conversely, some exams are simply pass or fail. In this case, you can focus your efforts on achieving a passing score.

▶ Match Your Abilities to Your Test

Now that you understand your strengths and weaknesses and you know what to expect of your test, you need to consider both factors to determine what material you need to study. First, look at the subject area or question type with which you have the most trouble. If you can expect to find questions of this type on your test, then this subject might be your first priority. But be sure to consider how much of the test will cover this material. For example, if there will only be a few questions out of one hundred that test your knowledge of a subject that is your weakest area, you might decide not to study this subject area at all. You might be better served by concentrating on solidifying your grasp of the main material covered on the exam.

The important thing to remember is that you want to maximize your time. You don't want to study material that you already know. And you don't want to study material that you don't need to know. You will make the best use of your time if you study the material that you know the least, but that you most need to know.

WHERE TO STUDY

The environment in which you choose to study can have a dramatic impact on how successful your studying is. If you choose to study in a noisy coffee shop at a small table with dim lighting, it might take you two hours to cover the same material you could read in an hour in the quiet of the library. That is an hour that you don't have to lose! However, for some people the noisy coffee shop is the ideal environment. You need to determine what type of study environment works for you.

▶ Consider Your Options

Your goal is to find a comfortable, secure place that is free from distractions. The place should also be convenient and conform to your schedule. For example, the library might be ideal in many respects. However, if it takes you an hour to get there and it closes soon after you arrive, you are not maximizing your study time.

For many people, studying at home is a good solution. Home is always open and you don't waste any time getting there, but it can have drawbacks. If you are trying to fit studying in between family obligations, you might find that working from home offers too many opportunities for distraction. Chores that have piled up, children or younger siblings who need your attention, or television that captures your interest, are just some of things that might interfere with studying at home. Maybe you have roommates who will draw your attention away from your studies. Studying at home is a good solution if you have a room that you can work in alone and away from any distractions.

If home is not a good environment for quiet study, the library, a reading room, or a coffee shop are places you can consider. Be sure to pick a place that is relatively quiet and which provides enough workspace for your needs.

▶ *Noise*

Everyone has his or her own tolerance for noise. Some people need absolute silence to concentrate, while others will be distracted without some sort of background noise. Classical music can be soothing and might help you relax as you study. In fact, studies have shown that listening to classical music actually enhances math performance. If you think you work better with music or the television on, you should be sure that you are not paying attention to what is on in the background. Try reading a chapter or doing some problems in silence, then try the same amount of work with noise. Which noise level allowed you to work the fastest?

▶ *Light*

You will need to have enough light to read comfortably. Light that is too dim will strain your eyes and make you drowsy. Too bright and you will be uncomfortable and tense. Experts suggest that the best light for reading comes from behind and falls over your shoulder. Make sure your light source falls on your book and does not shine in your eyes.

▶ *Comfort*

Your study place should be comfortable and conducive to work. While your bed might be comfortable, studying in bed is probably more conducive to sleeping than concentrated learning. You will need a comfortable chair that offers good back support and a work surface—a desk or table—that gives you enough space for your books and other supplies. Ideally, the temperature should be a happy medium between too warm and too cold. A stuffy room will make you sleepy and a cold room is simply uncomfortable. If you are studying outside your home, you may not be able to control the temperature, but you can dress appropriately. For example, bring along an extra sweater if your local library is skimpy with the heat.

▶ *A Little Help*

When you have settled on a place to study, you will need to enlist the help of your family and friends—especially if you are working at home. Be sure they know that when you go to your room and close the door that you do not want to be disturbed. If your friends all go to the same coffee shop where you plan to study, you will also need to ask them to respect your study place. The cooperation of your family and friends will eliminate one of the greatest potential distractions.

WHEN TO STUDY

Finding the time in your busy schedule may seem like the greatest hurdle in making your "just in time" study plan, but you probably have more time available than you think. It just takes a little planning and some creativity.

▶ Analyze Your Schedule

Your first step in finding time to study is to map out your day-to-day schedule—in detail. Mark a piece of paper in fifteen-minute intervals from the time you get up to the time you generally go to bed. Fill in each fifteen-minute interval. For example, if you work from nine to five, do not simply block that time off as unavailable for study. Write down your daily routine at work and see when you might have some time to study. Lunch is an obvious time, but there may be other down times in your workday when you can squeeze in a short study session.

You will want to set aside a stretch of time when you plan to study in your designated study place. But, you can also be creative and find ways to study for short bursts during your normal routine. For example, if you spend an hour at the gym on the stationary bike, you can read while you cycle. Or, you can review flashcards on your bus ride. If you drive to work, you could record some study material on a tape or CD. You could also listen to this tape while you walk the dog.

When you look at your schedule closely, you will probably find you have more time than you thought. However, if you still don't have the time you need, you should rethink your routine. Can you ask someone to take on a greater share of the household chores for the few weeks you need to get ready for your test? Is there some activity that you can forgo for the next few weeks? If you normally go to the gym six days a week for an hour and a half, cut down to three days a week for forty-five minutes. You will add over six and a half hours to your schedule without completely abandoning your fitness routine. Remember, any changes you make to your schedule are short-term and are a small sacrifice, once you consider your goal.

▶ Time Strategies

Now you know that when you have time available, you need to use that time to your best advantage. You will probably find that you can set aside one block of time during the day during which you will do the bulk of your studying. Use this time to learn new material or take a practice quiz and review your answers. Use the small spurts of time you have found in your schedule to review with flashcards, cheat sheets, and other tools.

Also, consider your learning style and body rhythm when you make your schedule. Does it take you some time to get into material? If so, you should build a schedule with longer blocks of time. Do you have a short attention span? Then you will do better with a schedule of several shorter study periods. No matter your style, avoid extremes. Neither very long study sessions nor very short (except for quick reviews) sessions are an efficient use of time. Whether you are a morning person or a night owl, plan to study when you are most energetic and alert.

Make sure your schedule allows for adequate rest and study breaks. Skipping sleep is not a good way to find time in your schedule. Not only will you be tired when you study, but also you will be sleep deprived by the time of the test. A sleep-deprived test-taker is more likely to make careless mistakes, lose energy and focus, and become stressed-out by the testing environment. If you plan to do most of your studying in one block of time, say four hours, be sure you leave time to take a study break. Experts have shown that students are more likely to retain material if they take some time to digest it. A five- or ten-minute break to stretch your legs or eat a snack will revive you and give your brain time to absorb what you have learned.

HOW TO STUDY

How you study is just as important as how long—especially if your time is limited. You will need to be in a good physical and mental state. And you will need to use the right tools for the job. You will also need to understand your learning style so that you can select the best study method. And, perhaps most important, you will need methods that will help you to remember not to memorize the material. All these techniques—using the right tools and methods—will help you make the most of your study time.

▶ Sleep Well, Eat Right, and Relax

Does your idea of studying hard include images of staying up into the wee hours and living on fast food and caffeine until the big test? Even though it may seem like you are working hard when you study around the clock and put aside good eating habits in order to save time, you are not working efficiently. If you have ever pulled an all-nighter you know that by four in the morning you can find yourself reading the same page several times without understanding a word. Adequate rest and good nutrition will allow you to be focused and energetic so you can get more work done in less time.

Most people need about eight hours of sleep a night. Do not sacrifice sleep in order to make time to study. Hunger can be a distraction, so don't skip meals. Eat three nutritious meals a day, and keep healthy snacks on hand during a long study session. The key word is healthy. Sugary snacks might make you feel energized in the short term, but that sugar rush is followed by a crash that will leave you feeling depleted. Caffeine can have a similar effect. A little caffeine—a morning cup of coffee, for example—can give you a boost, but too much caffeine will make you feel jittery and tense. Tension can affect your ability to concentrate.

Being over-caffeinated is not the only potential source of tension. Pre-exam anxiety can also get in the way of effective studying. If your anxiety about the upcoming test is getting the better of you, try these simple relaxation techniques:

- **Breathe!** Sounds simple, and it is. Taking long, deep breaths can drain the tension from your body. Place one hand on your stomach and the other on your chest. Sit up straight. Inhale deeply through your nose and feel your stomach inflate. Your chest should remain still. Exhale slowly through your mouth and feel your stomach deflate. It is the slow exhalation that helps you relax, so make sure you take your time releasing your breath. Pausing during a study session to take three deep breaths is a quick way to clear your mind and body of tension so that you can better focus on your work.
- **Tense and relax your muscles.** You may not even notice it, but as anxiety mounts, your muscles tense. You may tense your neck and shoulders, your toes, or your jaw. This tension can interfere with your concentration. Release the tension held in your muscles by purposefully tensing then relaxing each muscle. Work from your toes to your head systematically.
- **Visualize a soothing place.** Taking a break to mentally visit a place that you find relaxing can be invigorating. Close your eyes and conjure up the sights, smells, and sounds of your favorite place. Really try to feel like you are there for five uninterrupted minutes and you will return from your mini vacation ready to study.

▶ The Right Tools for the Job

If you follow the steps above, you will have a rested, energized, and relaxed brain—the most important tool you need to prepare for your exam. But there are other tools that you will need to make your study session the most productive. Be sure that you have all the supplies you need on hand before you sit down to study. To help make studying more pleasant, select supplies that you enjoy using. Here is a list of supplies that you will need:

- a notebook or legal pad dedicated to studying for your test
- graph paper
- pencils

- pencil sharpener
- highlighter
- index or other note cards
- paper clips or sticky note pads for marking pages
- a calendar or personal digital assistant (which you will use to keep track of your study plan)
- a calculator

▶ Break It Down

You may be feeling overwhelmed by the amount of material you have to cover in a short time. This seeming mountain of work can generate anxiety and even cause you to procrastinate further. Breaking down the work into manageable chunks will help you plan your studying and motivate you to get started. It will also help you organize the material in your mind. When you begin to see the large topic as smaller units of information that are connected, you will develop a deeper understanding of the subject. You will also use these small chunks of information to build your study plan. This will give you specific tasks to accomplish each day, rather than simply having time set aside to study for the test.

For example, if you have difficulty with geometry, you could study a different geometry topic each day for a week: On Monday, practice working with surface area; on Tuesday, work on Pythagorean theorem problems; on Wednesday, try coordinate geometry; and so on. "Learn geometry" might seem like an overwhelming task, but if you divide the work into smaller pieces, you will find that your understanding of geometry improves with practice and patience.

▶ Your Learning Style

Learning is not the same for everyone. People absorb information in different ways. Understanding how you learn will help you develop the most effective study plan for your learning style. Experts have identified three main types of learners: visual, auditory, and kinesthetic. Most people use a combination of all three learning styles, but one style might be more dominant. Here are some questions that will help you identify your dominant learning style:

1. If you have to remember an unusual word, you most likely
   a. picture the word in your mind.
   b. repeat the word aloud several times.
   c. trace out the letters with your finger.

2. When you meet new people, you remember them mostly by
  a. their actions and mannerisms.
  b. their names (faces are hard to remember).
  c. their faces (names are hard to remember).

3. In class you like to
  a. take notes, even if you don't reread them.
  b. listen intently to every word.
  c. sit up close and watch the instructor.

A visual learner would answer **a**, **c**, and **c**. An auditory learner would answer **b**, **b**, and **b**. A kinesthetic learner would answer **c**, **a**, and **a**.

Visual learners like to read and are often good spellers. When visual learners study, they often benefit from graphic organizers such as charts and graphs. Flashcards often appeal to them and help them learn, especially if they use colored markers, which will help them form images in their minds as they learn words or concepts.

Auditory learners, by contrast, like oral directions and may find written materials confusing or boring. They often talk to themselves and may even whisper aloud when they read. They also like being read aloud to. Auditory learners will benefit from saying things aloud as they study and by making tapes for themselves and listening to them later. Oral repetition is also an important study tool. Making up rhymes or other oral mnemonic devices will also help them study, and they may like to listen to music as they work.

Kinesthetic learners like to stay on the move. They often find it difficult to sit still for a long time and will often tap their feet and gesticulate a lot while speaking. They tend to learn best by doing rather than observing. Kinesthetic learners may want to walk around as they practice what they are learning, because using their bodies helps them remember things. Taking notes is an important way of reinforcing knowledge for the kinesthetic learner, as is making flashcards.

It is important to remember that most people learn in a mixture of styles, although they may have a distinct preference for one style over the others. Determine which is your dominant style, but be open to strategies for all types of learners.

▶ Remember–Don't Memorize

You need to use study methods that go beyond rote memorization to genuine comprehension in order to be fully prepared for your test. Using study methods that suit your learning style will help you to *really* learn the material you need to know for the test. One of the most important learning strategies is to be an active reader. Interact with what you are reading by asking questions, making notes, and marking passages instead of simply reading the words on the page. Choose methods of interacting with the text that match your dominant learning style.

- **Ask questions.** When you read a lesson, ask questions such as, "What is the main idea of this section?" Asking yourself questions will test your comprehension of the material. You are also putting the information into your own words, which will help you remember what you have learned. This can be especially helpful when you are learning math techniques. Putting concepts into your own words helps you to understand these processes more clearly.
- **Make notes.** Making notes as you read is another way for you to identify key concepts and to put the material into your own words. Writing down important ideas and mathematical formulas can also help you memorize them.
- **Highlight.** Using a highlighter is another way to interact with what you are reading. Be sure you are not just coloring, but highlighting key concepts that you can return to when you review.
- **Read aloud.** Especially for the auditory learner, reading aloud can help aid in comprehension. Hearing mathematical information and formulas read aloud can clarify their meanings for you.
- **Make connections.** Try to relate what you are reading to things you already know or to a real-world example. It might be helpful, for example, to make up a word problem, or draw a diagram or table, to clarify your understanding of what a problem is asking you to do.

Reading actively is probably the most important way to use your study time effectively. If you spend an hour passively reading and retaining little of what you have read, you have wasted that hour. If you take an hour and a half to actively read the same information, that is time well spent. However, you will not only be learning new material; you will also need methods to review what you have learned:

- **Flashcards.** Just making the cards alone is a way of engaging with the material. You have to identify key concepts, words, or important information and write them down. Then, when you have made a stack of cards, you have a portable review system. Flashcards are perfect for studying with a friend and for studying on the go.
- **Mnemonics.** These catchy rhymes, songs, and acronyms are tools that help us remember information. Some familiar mnemonics are "*i* before *e* except after *c*" or ROY G. BIV, which stands for Red Orange Yellow Green Blue Indigo Violet—the colors of the rain-

bow. Developing your own mnemonics will help you make a personal connection with the material and help you recall it during your test. Mnemonics are also useful when you personalize your "cheat sheet."

- **Personalized cheat sheet.** Of course, you aren't really going to cheat, but take the Formula Cheat Sheet found on pages ix–x and add to it. Or, highlight the formulas you really need and don't yet know well. This will help them to stand out more than the ones you already know. You can then use the sheet to review—it's perfect for studying on the go.

- **Outlines and Maps.** If you have pages of notes from your active reading, you can create an outline or map of your notes to review. Both tools help you organize and synthesize the material. Most students are familiar with creating outlines using hierarchical headings, but maps may be less familiar. To make a map, write down the main point, idea, or topic under consideration in the middle of a clean piece of paper. Draw a circle around this main topic. Next, draw branches out from that center circle on which to record subtopics and details. Create as many branches as you need—or as many as will fit on your sheet of paper.

▶ Studying with Others

Studying in a group or with another person can be a great motivator. It can also be a distraction, as it can be easy to wander off the subject at hand and on to more interesting subjects such as last night's game, or some juicy gossip. The key is to choose your study partners well and to have a plan for the study session that will keep you on track.

There are definite advantages to studying with others:

- **Motivation.** If you commit to working with someone else, you are more likely to follow through. Also, you may be motivated by some friendly competition.

- **Solidarity.** You can draw encouragement from your fellow test-takers and you won't feel alone in your efforts. This companionship can help reduce test anxiety.

- **Shared expertise.** As you learned from your practice questions, you have certain strengths and weaknesses in the subject. If you can find a study partner with the opposite strengths and weaknesses, you can each benefit from your partner's strengths. Not only will you get help, but also you will build your confidence for the upcoming test by offering *your* expertise.

There are also some disadvantages to studying with others:

- **Stress of competition.** Some study partners can be overly competitive, always trying to prove that they are better in the subject than you. This can lead to stress and sap your confidence. Be wary of the overly competitive study partner.
- **Too much fun.** If you usually hate studying, but really look forward to getting together with your best friend to study, it may be because you spend more time socializing than studying. Sometimes it is better to study with an acquaintance who is well-matched with your study needs and with whom you are more likely to stay on task.
- **Time and convenience.** Organizing a study group can take time. If you are spending a lot of time making phone calls and sending e-mails trying to get your study group together, or if you have to travel a distance to meet up with your study partner, this may not be an efficient strategy.

Weigh the pros and cons of studying with others to decide if this is a good strategy for you.

JUST THE FACTS . . . JUST IN TIME

You have thought about the what, where, when, and how; now you need to put all four factors together to build your study plan. Your study plan should be as detailed and specific as possible. When you have created your study plan, you then need to follow through.

▶ Building a Study Plan

You will need a daily planner, a calendar with space to write, or a personal digital assistant to build your plan. You have already determined the time you have free for study. Now, you need to fill in the details. You have also figured out what you need to study, and have broken the material down into smaller chunks. Assign one chunk of material to each of the longer study sessions you have planned. You may need to combine some chunks or add some review sessions depending on the number of long study sessions you have planned in your schedule.

You can also plan how to study in your schedule. For example, you might write for Monday 6:00 P.M. to 9:00 P.M.: Read Chapter 4, make notes, map notes, and create set of flashcards. Then for Tuesday 8:30 A.M. to 9:00 A.M. (your commute time): study Chapter 4 flashcards. The key to a successful study plan is to be as detailed as possible.

▶ Staying on Track

Bear in mind that nothing goes exactly as planned. You may need to stay late at work, you may get a nasty cold, soccer practice may go late, or your child might need to go to the doctor: any number of things can happen to your well-thought-out study plan—and some of them probably will. You will need strategies for coping with life's little surprises.

The most important thing to remember when you get off track is not to panic or throw in the towel. You can adjust your schedule to make up the lost time. You may need to reconsider some of your other commitments and see if you can borrow some time for studying. Or, you may need to forgo one of your planned review sessions to learn new material. You can always find a few extra minutes here and there for your review.

▶ Minimizing Distractions

There are some distractions, such as getting sick, that are unavoidable. Many others can be minimized. There are the obvious distractions such as socializing, television, and the telephone. There are also less amusing distractions such as anxiety and fear. They can all eat up your time and throw off your study plan. The good news is you can do a lot to keep these distractions at bay.

- **Enlist the help of your friends and family.** Just as you have asked your friends and family to respect your study space, you can also ask them to respect your study time. Make sure they know how important this test is to you. They will then understand that you don't want to be disturbed during study time, and will do what they can to help you stick to your plan.
- **Keep the television off.** If you know that you have the tendency to get pulled into watching TV, don't turn it on even *before* you plan to study. This way you won't be tempted to push back your study time to see how a program ends or see what's coming up next.
- **Turn off your cell phone and the ringer on your home phone.** This way you won't eat up your study time answering phone calls—even a five-minute call can cause you to lose focus and waste precious time.
- **Use the relaxation techniques discussed earlier in the chapter if you find yourself becoming anxious while you study.** Breathe, tense and relax your muscles, or visualize a soothing place.
- **Banish negative thoughts.** Negative thoughts—such as, "I'll never get through what I planned to study tonight," "I'm so mad all my friends are at the movies and I'm stuck here studying," "Maybe I'll just study for an hour instead of two so I can watch the season finale

of my favorite show"—interfere with your ability to study effectively. Sometimes just noticing your negative thoughts is enough to conquer them. Simply answer your negative thought with something positive—"If I study the full two hours, I can watch the tape of my show," "I want to study because I want to do well on the test so I can . . ." and so on.

▶ Staying Motivated

You can also get off track because your motivation wanes. You may have built a rock-solid study plan and set aside every evening from 6:00 to 9:00 to study. And then your favorite team makes it to the playoffs. Your study plan suddenly clashes with a very compelling distraction. Or, you may simply be tired from a long day at work or school or from taking care of your family and feel like you don't have the energy for three hours of concentrated study. Here are some strategies to help keep you motivated:

- **Visualization.** Remind yourself of what you will gain from doing well on the test. Take some time to visualize how your life will be positively changed if you accomplish your goal. Do not, however, spend time visualizing how awful your life will be if you fail. Positive visualization is a much more powerful motivator than negative imagery.
- **Rewards.** Rewards for staying on track can be a great motivator, especially for flagging enthusiasm. When you accomplish your study goal, perhaps watch your favorite TV program or have a special treat—whatever it is that will motivate you.
- **Positive feedback.** You can use your study plan to provide positive feedback. As you work toward the test date, look back at your plan and remind yourself of how much you have already accomplished. Your plan will provide a record of your steady progress as you move forward. You can also enlist the help of study partners, family, and friends to help you stay motivated. Let the people in your life know about your study plan and your progress. They are sure to applaud your efforts.

At the end of the day, *you* will be your prime motivator. The fact that you bought this book and have taken the time to create a well-thought-out study plan shows that you are committed to your goal. Now, all that is left is to go for it! Imagine yourself succeeding on your test and let the excitement of meeting your goal carry you forward.

2

Building Blocks of Geometry: Points, Lines, and Angles

The study of geometry begins with an understanding of the basic building blocks: points, lines, segments, rays, and angles. All geometric figures are a result of these simple figures. Begin reviewing geometry by taking this ten-question benchmark quiz. These questions are similar to the type of questions that you will find on important tests. When you are finished, check the answer key carefully to assess your results. Use this quiz to determine how much of your study time should be devoted to this chapter.

BENCHMARK QUIZ

Use the drawing below to answer questions 1 through 4.

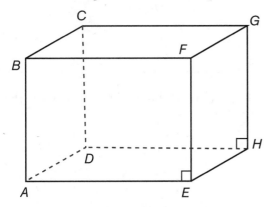

1. Which set of points is NOT coplanar?
   a. *A, E, F*, and *B*
   b. *A, B, C*, and *D*
   c. *A, B, C*, and *E*
   d. *A, E, H*, and *D*
   e. All the sets above are coplanar.

2. Which lines are skew lines?
   a. $\overleftrightarrow{BF}$ and $\overleftrightarrow{DH}$
   b. $\overleftrightarrow{EH}$ and $\overleftrightarrow{HG}$
   c. $\overleftrightarrow{AB}$ and $\overleftrightarrow{GH}$
   d. $\overleftrightarrow{BF}$ and $\overleftrightarrow{EH}$
   e. $\overleftrightarrow{EF}$ and $\overleftrightarrow{AB}$

3. Which lines are NOT parallel?
   a. $\overleftrightarrow{CD}$ and $\overleftrightarrow{AB}$
   b. $\overleftrightarrow{GH}$ and $\overleftrightarrow{AB}$
   c. $\overleftrightarrow{AE}$ and $\overleftrightarrow{AB}$
   d. $\overleftrightarrow{CG}$ and $\overleftrightarrow{AE}$
   e. $\overleftrightarrow{BF}$ and $\overleftrightarrow{DH}$

4. Which lines are perpendicular?
   a. $\overleftrightarrow{AB}$ and $\overleftrightarrow{FE}$
   b. $\overleftrightarrow{FG}$ and $\overleftrightarrow{AB}$
   c. $\overleftrightarrow{DH}$ and $\overleftrightarrow{HG}$
   d. $\overleftrightarrow{HE}$ and $\overleftrightarrow{CB}$
   e. $\overleftrightarrow{GH}$ and $\overleftrightarrow{AB}$

5. If the measure of $\overline{AC}$ is 28 inches, the measure of $\overline{AB}$ is 2x + 4, and point B is the midpoint of $\overline{AC}$, what is the value of x?
   a. 5 inches
   b. 12 inches
   c. 14 inches
   d. 7 inches
   e. 3.5 inches

6. If the measure of ∠WYZ = 35° and $\overrightarrow{YW}$ bisects ∠XYZ, what is the measure of ∠XYW?

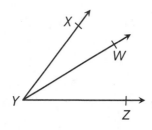

   a. 70°
   b. 105°
   c. 17.5°
   d. 35°
   e. 350°

7. If $\overline{YZ}$ is the perpendicular bisector of $\overline{AB}$, and the segments intersect at point W, which of the following would ALWAYS be true?
   a. The measure of $\overline{AW}$ is one-half the measure of $\overleftrightarrow{WB}$.
   b. ∠AWZ is 90°.
   c. ∠AWY is acute.
   d. The measure of $\overline{YW}$ is one half the measure of $\overline{YZ}$.
   e. The sum of ∠AWY and ∠BWY is 90°.

8. Given the figure below, which of the following is true?

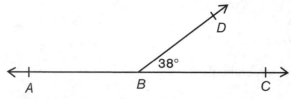

   a. $\overrightarrow{AB}$ is perpendicular to $\overrightarrow{AD}$.
   b. $\overrightarrow{BD}$ is parallel to $\overrightarrow{BC}$.
   c. Points A, B, C, and D are collinear.
   d. The sum of the measure of ∠ABD and ∠CBD add up to 90°.
   e. The sum of the measure of ∠ABD and ∠CBD add up to 180°.

9. Which is the classification of the angle?

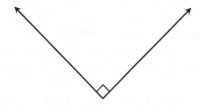

   a. right acute
   b. right
   c. obtuse
   d. right obtuse
   e. acute

10. Given that the lines l and m are parallel, $\overline{AC}$ is perpendicular to line l, and $\overline{BD}$ is perpendicular to line l, which statement or statements are FALSE?

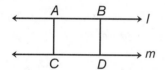

   a. ∠ACD is congruent to ∠BCD.
   b. The measure of $\overline{AC}$ is equal to the measure of $\overline{BD}$.
   c. ∠BDC is 90°.
   d. ∠ACD is 90°.
   e. $\overline{BD}$ is perpendicular to line m.

BENCHMARK QUIZ SOLUTIONS

How did you do on the subject of the basic building blocks of geometry? Check your answers here, and then analyze your results to figure out your plan of attack to master these topics.

   1. c. Answer choices **a**, **b**, and **d** are coplanar points. Choice **c**, points A, B, C, and E, are NOT coplanar. Points A, E, C, and G lie on a plane that cuts through the solid on a diagonal. Points A, B, C, and D lie on the plane that is a side of the solid. Points A, E, H, and D lie on a plane that is the base of the solid.

2. d. Skew lines are two non-parallel lines in different planes that do not intersect. $\overleftrightarrow{BF}$ and $\overleftrightarrow{EH}$ are not contained in the same plane, so they are not parallel and will never intersect. All the other line pairs are either parallel in the same plane, or intersect in the same plane. Some of the pairs are in planes that are diagonal through the solid, and some pairs are in planes that are sides of the solid.

3. c. $\overleftrightarrow{AE}$ and $\overleftrightarrow{AB}$ intersect, and are actually perpendicular lines. Parallel lines do not intersect, and lie in the same plane. All of the other pairs of lines are parallel in a plane. Some of the pairs are in planes that are diagonal through the solid, and some pairs are in planes that are sides of the solid.

4. c. Perpendicular lines form a right angle. Although the angle does not look like a right angle, it is 90°, as it forms the bottom right angle of the rectangular solid, and it is labeled with the small square in the angle to indicate 90°. All of the other pair choices are either parallel or skew lines.

5. a. Since no picture is given, sketch the segment:

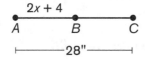

Given the information and the figure, set up an equation to solve for x. Since B is the midpoint, the measure of $\overline{AB}$ plus the measure of $\overline{BC}$ equals 28. The equation is:

$2x + 4 + 2x + 4 = 28$	
$4x + 8 = 28$	Combine like terms.
$4x + 8 - 8 = 28 - 8$	Subtract 8 from both sides.
$\dfrac{4x}{4} = \dfrac{20}{4}$	Combine like terms, and divide both sides by 4.

$x = 5$ inches

6. d. Since $\overrightarrow{YW}$ bisects $\angle XYZ$, the measure of $\angle XYW$ is equal to the measure of $\angle WYZ$, which is given as 35°.

7. b. Since no picture is given, sketch the segments:

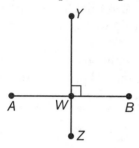

Perpendicular segments form 90° angles and ∠AWZ is formed by the perpendicular segments. Note that for choice **d**, it is never stated that $\overline{AB}$ intersects $\overline{YZ}$, so you cannot assume that it bisects.

8. e. The angles form a straight line, and all straight lines form a straight angle, which is equal to 180°.

9. b. The angle is a right angle, as indicated by the small square in the interior of the angle.

10. a. ∠ACD is NOT congruent to ∠BCD. Remember that the middle letter is the vertex of the angle. ∠ACD is a right angle, formed by the perpendicular line and segment. ∠BCD is an acute angle. It is not shown in the figure, but it is definitely less than 90°. Draw in the angle with a dotted line as shown:

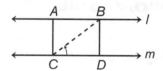

BENCHMARK QUIZ RESULTS

If you answered 8–10 questions correctly, you have a good grasp of the basic building blocks of geometry. Read over the chapter, concentrating on the areas where you were weak. Then, proceed to the quiz at the end of this chapter for additional confirmation of your success.

If you answered 4–7 questions correctly, you need to refresh yourself on these topics. Carefully read through the lesson in this chapter for review and skill building. Pay attention to the sidebars that refer you to more in-depth practice, hints, and shortcuts. Work through the quiz at the end of the chapter to check your progress.

If you answered 1–3 questions correctly, you need help and clarification on the topics in this section. First, carefully read this chapter and concentrate on the sidebars and visual aids that will help with comprehension. Perhaps you learned this information and forgot; take the time now to refresh your skills and improve your knowledge. Go to the suggested website in the Extra Help sidebar in this chapter, and do extended practice. You may also want to refer to *Geometry Success in 20 Minutes a Day*, Lessons 1, 2, 3, and 4, published by LearningExpress.

JUST IN TIME LESSON—BASIC BUILDING BLOCKS OF GEOMETRY

Topics in this chapter include:

- Geometry Vocabulary: Point, Line, and Plane
- Types of Points
- Geometry Vocabulary: Ray and Segment
- Types of Lines and Planes
- Angles and Angle Classification
- Geometry Vocabulary: Congruent Figures, Midpoint, and Bisector
- Using Algebra to Solve Geometric Problems
- Perpendicular Lines and Perpendicular Bisectors

GEOMETRY VOCABULARY—POINT, LINE, AND PLANE

All of geometry starts with three basic constructs.

 GLOSSARY

POINT a position in space. A point has no length, width, or height.
LINE an infinite collection of points that has length, but no width or thickness
PLANE an infinite collection of points that has length, and width but no thickness

A point can be pictured by a dot and is labeled with a capital letter.

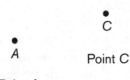

Point C

Point A

Lines in geometry are straight lines unless otherwise described. Any two points determine exactly one unique line. Two points on the line, in either order, name a line. The following graphic is a picture of $\overleftrightarrow{AB}$ or $\overleftrightarrow{BA}$.

There can be many names for a line. If we label another point on the line, the line on the following graphic can be labeled $\overleftrightarrow{AB}$, $\overleftrightarrow{BA}$, $\overleftrightarrow{CA}$, $\overleftrightarrow{AC}$, $\overleftrightarrow{CB}$, or $\overleftrightarrow{BC}$.

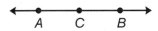

A line, as indicated by the arrows on each end, extends forever in either direction.

A plane is a flat surface. It can be envisioned as a sheet of paper that extends in all four directions forever. Any three non-collinear points determine a plane. Any three non-collinear points on its surface name a plane.

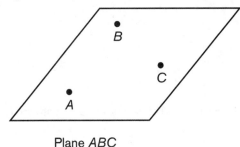

Plane *ABC*

TYPES OF POINTS

GLOSSARY

COLLINEAR POINTS three or more points that lie on the same line
NON-COLLINEAR POINTS three or more points that do not lie on the same line

Example:

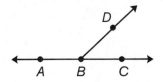

Points *A*, *B*, and *C* are collinear. Points *B*, *C*, and *D* are non-collinear.

GEOMETRY VOCABULARY: RAY AND SEGMENT

GLOSSARY

RAY an endpoint P on a line and all the points on the line that lie on one side of P. There are an infinite number of points on a ray.

Rays are named with the endpoint and another point on the ray. The endpoint is always listed first. The following is a picture of $\overrightarrow{XY}$ with endpoint X.

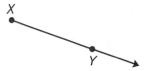

Note that when naming a ray, the endpoint must be named first. In the same figure, there is a line, $\overleftrightarrow{XY}$. It is made up of two different rays. $\overrightarrow{XY}$ is different than $\overrightarrow{YX}$. If we label another point on $\overrightarrow{XY}$, as shown in the following figure, we can name the same ray, that is $\overrightarrow{XY}$, with another name, $\overrightarrow{XZ}$.

GLOSSARY

SEGMENT two endpoints, such as R and T, and all of the points that lie between R and T

Following is a picture of $\overline{RT}$ with endpoints R and T.

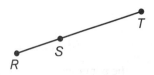

This segment can be named either $\overline{RT}$ or $\overline{TR}$. A segment has a definite length that can be measured with a ruler. There are, however, an infinite number of points on $\overline{RT}$. If we add another point, point S, to $\overline{RT}$, we create two new distinct segments, $\overline{RS}$ or $\overline{SR}$, and $\overline{ST}$ or $\overline{TS}$.

Example:
Given the figure:

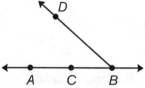

There is: One line, named $\overleftrightarrow{AB}$ or $\overleftrightarrow{BA}$, $\overleftrightarrow{AC}$ or $\overleftrightarrow{CA}$, $\overleftrightarrow{BC}$ or $\overleftrightarrow{CB}$
Collinear points A, B, and C
Non-collinear points A, B, and D
Four segments, named $\overline{AB}$ or $\overline{BA}$, $\overline{BC}$ or $\overline{CB}$, $\overline{AC}$ or $\overline{CA}$, $\overline{BD}$ or $\overline{DB}$
Five rays, named $\overrightarrow{BA}$ or $\overrightarrow{BC}$, $\overrightarrow{BD}$, $\overrightarrow{AB}$ or $\overrightarrow{AC}$, $\overrightarrow{CB}$, or $\overrightarrow{CA}$

SHORTCUT

Remember that all lines, segments, rays, and planes have an infinite number of points. There may be only a few points named, but there exists an infinite number of points on these figures.

TYPES OF LINES AND PLANES

 GLOSSARY

PARALLEL LINES lines in the same plane that do not intersect

The following is a picture of parallel lines. Often, arrowheads are shown on a pair of parallel lines.

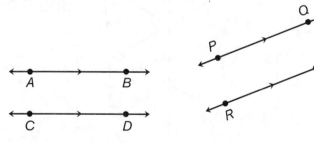

Parallel is denoted by the symbol ‖ so in the preceding figure, $\overleftrightarrow{AB} \parallel \overleftrightarrow{CD}$ and $\overleftrightarrow{PQ} \parallel \overleftrightarrow{RS}$. Segments and rays can also be parallel. Likewise, there are parallel planes. When two distinct parallel planes are pictured, the third dimension is introduced, and it becomes more difficult to envision and depict on a page.

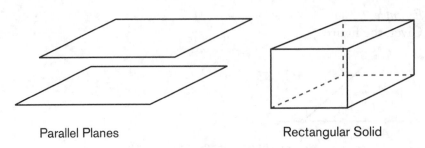

Parallel Planes   Rectangular Solid

In the rectangular solid above, there are several pairs of parallel planes, namely the top and bottom, the left and right, and the front and back planes.

RULE BOOK

Any two distinct lines in the same plane are either parallel or intersecting. If they intersect (meet), they meet in exactly one point.
Two distinct planes can be parallel or intersecting. If they intersect, they meet in exactly one line:

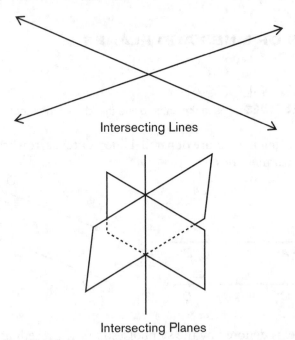

Intersecting Lines

Intersecting Planes

 GLOSSARY

SKEW LINES two non-parallel lines that do not intersect. These two lines are not in the same plane.

Skew lines are envisioned in three dimensions. Segments and rays can also be skew. In the following figure, $\overline{HI}$ and $\overline{NM}$, $\overline{HK}$ and $\overline{JN}$, and also $\overline{NO}$ and $\overline{IM}$ are examples of pairs of segments that are skew.

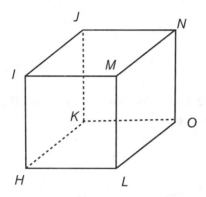

ANGLES AND ANGLE CLASSIFICATION

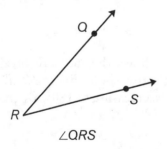 GLOSSARY

ANGLE a geometric figure formed by two rays with the same endpoint

Angles are also formed by intersecting lines, and by two segments, with one of the endpoints the same. In an angle, the common endpoint is called the *vertex* and the *rays*, or *segments*, are called the *sides*. The following is a picture of $\angle QRS$, with vertex R and sides $\overrightarrow{RQ}$ and $\overrightarrow{RS}$.

$\angle QRS$

If there is no other angle at vertex R, the angle can be named $\angle R$. Otherwise, all three points must be named, with the vertex point named in the middle, such as $\angle QRS$ or $\angle SRQ$. Sometimes the interior is labeled with a number to make reference to the angle easier, as shown in the following picture.

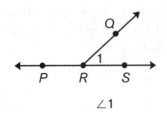

∠1

RULE BOOK

When naming an angle with three points, the vertex point MUST be the middle point listed. For example:

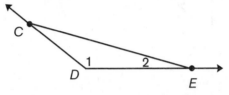

The angle labeled ∠1 is ∠CDE or ∠EDC, but ∠2 is ∠CED or ∠DEC. Notice that the same points, C, D, and E, together with the segments and rays shown, create several distinct angles.

An angle has a definite size, measured by a protractor. Angle size can be envisioned as a vertex and one side of the angle on a line, and the circular distance to reach the other side determines the measure. The units of angle measured are called *degrees*.

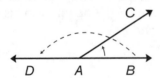

In the figure, the solid curved arrow shows the measure of ∠BAC, denoted by m∠BAC. The dotted curved arrow shows the m∠BAD. The m∠BAD, which is the straight line $\overleftrightarrow{DB}$, has a measure of 180°. Degrees are denoted by the ° symbol.

SHORTCUT

If there is a figure where two angles together form a straight line, the sum of their measure is 180°.

RULE BOOK

Angles are classified according to their measure:

ACUTE ANGLES are angles that measure greater than 0° and less than 90°.
RIGHT ANGLES are angles that measure 90°.
OBTUSE ANGLES are angles that measure greater than 90° and less than 180°.
STRAIGHT ANGLES are angles that measure 180°.
REFLEX ANGLES are angles that measure greater than 180°.

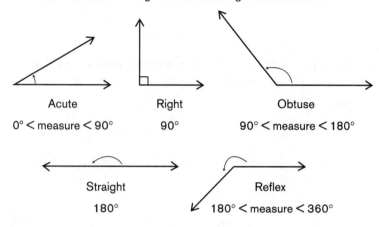

Acute	Right	Obtuse
0° < measure < 90°	90°	90° < measure < 180°

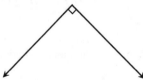

Straight   Reflex
180°   180° < measure < 360°

Example:
What is the classification of the following angle?

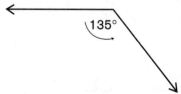

The above angle is exactly 90°. The classification is a right angle.

Example:
What is the classification of the following angle?

This angle is greater than 90° and less than 180°, so the classification is obtuse.

SHORTCUT

If there is a figure of an angle with the small box drawn in the interior, then the angle measure is 90°, and the angle is a right angle. If there is a figure where two angles together form an angle with the small box in the interior, the sum of their measures is 90°.

GEOMETRY VOCABULARY—CONGRUENT FIGURES, MIDPOINT, AND BISECTOR

GLOSSARY

CONGRUENT FIGURES have the same size and shape. Any segments or angles with the same size are congruent. The symbol for congruent is ≅.

Example:

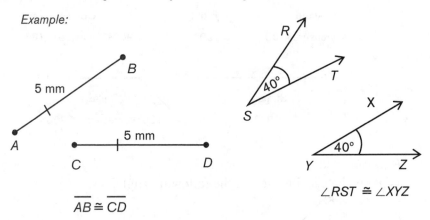

$$\overline{AB} \cong \overline{CD}$$

∠RST ≅ ∠XYZ

Congruent segments are marked with slashes and congruent angles are marked with arcs.

MIDPOINT of a segment is a point, such as Q, that divides a segment into two congruent segments

Example:

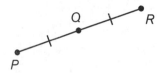

In this example, Q is the midpoint of $\overline{PR}$, therefore $\overline{PQ} \cong \overline{QR}$.

BISECTOR a line, segment, or ray that divides a figure into two congruent figures.

Example:

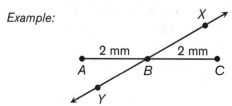

In this example, $\overrightarrow{XY}$ bisects $\overline{AC}$, since $\overline{AB} \cong \overline{BC}$.

Example:

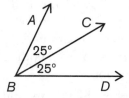

If $\overrightarrow{BC}$ bisects $\angle ABD$ and m$\angle ABD = 50°$, then m$\angle ABC =$ m$\angle CBD$ and each angle measures 25°.

USING ALGEBRA TO SOLVE GEOMETRIC PROBLEMS

Often, algebra is combined with the concepts of midpoint and bisector to solve problems.

Example:
Given the figure below, if the measure of $\overline{XZ}$ = 42 inches and Y is the midpoint of $\overline{XZ}$, what is the value of x?

To solve this, remember that the two smaller segments are congruent since point Y is the midpoint.

$6x + 3 + 6x + 3 = 42$	Set up an equation to solve the problem.
$12x + 6 = 42$	Combine like terms.
$12x + 6 - 6 = 42 - 6$	Subtract 6 from both sides of the equation.
$\frac{12x}{12} = \frac{36}{12}$	Combine like terms, and divide both sides by 12.
$x = 3$ inches	

Example:
Given the figure below, if $\overrightarrow{BD}$ bisects $\angle ABC$, m$\angle CBD = 67.5°$, and $\angle CBA = 12x + 15$, what is the value of x?

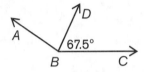

$\angle CBD$ is one half the measure of $\angle CBA$ since $\overrightarrow{BD}$ bisects $\angle ABC$. Therefore, the measure of $\angle ABC = 2 \times 67.5$, which is equal to 135°.

$12x + 15 = 135$	Set up an equation.
$12x + 15 - 15 = 135 - 15$	Subtract 15 from both sides of the equation.
$\frac{12x}{12} = \frac{120}{12}$	Combine like terms, and divide both sides by 12.
$x = 10$	

Be careful when using algebra to solve a problem. Read carefully and understand whether the problem requires you to find the value of the variable, or the actual value of the measure of an angle or segment. If the measure of an angle or segment is required, first solve for the variable and then use this value to solve for the measure. In the last example, if the problem had asked for the measure of $\angle CBA$, you would substitute 10 for the variable x, to get $12x + 15 = 12 \times 10 + 15 = 120 + 15 = 135$.

PERPENDICULAR LINES AND PERPENDICULAR BISECTOR

GLOSSARY
PERPENDICULAR LINES are distinct intersecting lines that form four congruent angles, each of 90°. The symbol for perpendicular is $\perp$.

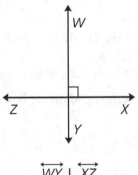

$\overleftrightarrow{WY} \perp \overleftrightarrow{XZ}$

PERPENDICULAR BISECTOR of a segment, is a segment, line, or ray that is perpendicular to the segment and also divides the segment into two congruent segments

In the following figure, the line $\overleftrightarrow{VZ}$ is a perpendicular bisector of segment $\overline{WY}$. So, by definition, $\angle WXV \cong \angle VXY \cong \angle YXZ \cong \angle ZXW$ and each has a measure of 90°. Notice that a small box is used to indicate perpendicular figures. This small box in the interior of any angle will indicate a 90° angle. Recall that the measure of $\angle YXW = 180°$ by definition; it forms a straight line. Since $\angle VXY \cong \angle WXV$, then their measures are equal. Together they measure 180°, so the measure of each separate angle is 90°. Segments and rays can also be perpendicular.

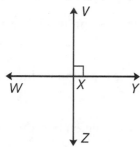

 RULE BOOK

If there are two parallel lines and a segment perpendicular to one of the lines, then the segment is also perpendicular to the other line. If there are two lines that are parallel, any perpendicular segments in between these lines, with endpoints on the lines, are congruent; they have the same measure.

Example:

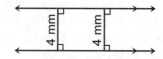

TIPS AND STRATEGIES

When working with the basic geometric figures, remember:

- Know the geometry vocabulary, as outlined under Glossary.
- Lines, planes, segments, and rays have an infinite number of points.
- Segments and angles can be measured.
- Any two lines in the same plane either intersect, or they are parallel.
- Two lines in the same plane that intersect meet at a single point.
- Two planes that intersect meet in a straight line.
- Skew lines are non-intersecting, are not in the same plane, and are not parallel.
- When naming an angle, the vertex point must be named in the middle.
- Angles are classified by their measure.
- A perpendicular bisector of a segment divides the segment into two congruent segments, and forms right angles.

When you are taking a multiple-choice test, remember this tip to improve your score:

- When using algebra to solve a geometry problem, be sure what the problem is asking you to find. Sometimes, the value of the variable is wanted, and other times the measure of a segment or angle is wanted.

EXTRA HELP

For further instruction and clarification on the basic geometric figures, visit the website *www.math.com*. On the left sidebar, click on *Geometry*. There is a series of interactive lessons, followed by a short interactive quiz. Another website is *http://library.thinkquest.org/2609/*. Click to enter the site, and then click on *Start the Lessons!* to take the quiz. You may also want to reference the book *Geometry Success in 20 Minutes a Day*, published by LearningExpress.

PRACTICE QUIZ

Following is additional practice on the basic geometric figures. Check to see if you have mastered these concepts.

Use this figure for questions 1 through 6.

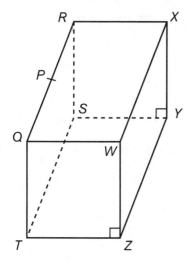

1. Which of the following statements is FALSE?
   a. Points R and Z lie on a line.
   b. Points Q, P, and R are collinear.
   c. Points R, W, and Z are collinear.
   d. Points Z, Y, and X are non-collinear.
   e. Points T, S, and W are non-collinear.

2. Which of the following statements is true?
   a. Points W, X, and S are coplanar.
   b. The intersection of plane WZY and plane TZY is point Z.
   c. Plane TQW is parallel to plane RXY.
   d. choice **a** and **c** only
   e. Choices **a**, **b**, and **c** are all true.

3. Which pair of lines are skew lines?
   a. $\overleftrightarrow{QT}$ and $\overleftrightarrow{RS}$
   b. $\overleftrightarrow{WX}$ and $\overleftrightarrow{SY}$
   c. $\overleftrightarrow{TS}$ and $\overleftrightarrow{SY}$
   d. $\overleftrightarrow{PW}$ and $\overleftrightarrow{RX}$
   e. None of the above are skew lines.

4. Which pair of lines is parallel?
   a. $\overleftrightarrow{QP}$ and $\overleftrightarrow{WX}$
   b. $\overleftrightarrow{QW}$ and $\overleftrightarrow{SY}$
   c. $\overleftrightarrow{QW}$ and $\overleftrightarrow{TZ}$
   d. choices **a** and **b** only
   e. All of the above pairs are parallel lines.

5. Which of the following statements is true?
   a. $\angle WZT$ measures 90°.
   b. $\angle ZYS$ is acute.
   c. $\angle WZT$ is right acute.
   d. choices **a** and **c** only
   e. choices **a**, **b**, and **c**

6. Which of the following sets of points is coplanar?
   a. points T, Z, R, and X
   b. points Q, R, S, and Z
   c. points W, X, Y, and Z
   d. choices **a** and **c** only
   e. choices **a**, **b**, and **c**

7. Given the figure below and the fact that $\overleftrightarrow{XY} \parallel \overleftrightarrow{AB}$, which of the following statements is ALWAYS true?

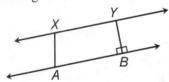

   a. $\overleftrightarrow{AB} \perp \overline{BY}$
   b. The measure of $\overline{BY}$ is equal to the measure of $\overline{XA}$.
   c. $\angle XYB \cong \angle BAY$
   d. $\overline{AX} \perp \overline{XY}$
   e. None of the above is always true.

8. What is the classification of the following angle?

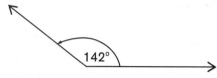

    a. straight obtuse
    b. obtuse
    c. right obtuse
    d. acute
    e. right acute

9. Given the following three angles, which of the following statements is true?

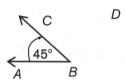

    a. $\angle DEF$ is the bisector of $\angle GHI$.
    b. All of the angles are congruent.
    c. $\angle ABC$ is an acute right angle.
    d. $\angle ABC \cong \angle DEF$
    e. All of the above statements are true.

10. Given the figure below and that $\overleftrightarrow{CE} \parallel \overleftrightarrow{DF}$, $\overline{EF} \perp \overleftrightarrow{CE}$, $\overline{CD} \perp \overleftrightarrow{DF}$, which of the following statements is true?

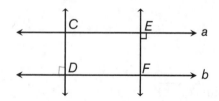

    a. $\overline{CD} \cong \overline{EF}$
    b. $\angle CDF$ is a right angle.
    c. $\overline{EF} \perp \overleftrightarrow{DF}$
    d. All of the above are true.
    e. None of the above is true.

11. If $\overleftrightarrow{CD} \perp \overleftrightarrow{CE}$, which of the following statements is true?
    a. Points C, D, and E are collinear.
    b. $\angle CED$ is a right angle.
    c. $\angle DCE$ is a right obtuse angle.
    d. All of the above are true.
    e. None of the above is true.

12. Given the following figure, which of the statements is true?

    a. $\overleftrightarrow{TR}$, together with $\overrightarrow{TU}$, form a 100° angle.
    b. $\overleftrightarrow{RS}$ and $\overleftrightarrow{SR}$ name the exact same set of points.
    c. There are exactly four points on $\overrightarrow{RU}$.
    d. There are exactly three points on $\overrightarrow{TR}$.
    e. $\overrightarrow{TS}$ and $\overrightarrow{TR}$ name the exact same set of points.

13. Given the following figure, which of the statements is true?

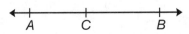

    a. You can measure the length of line $\overleftrightarrow{AB}$.
    b. $\overleftrightarrow{AB}$ contains the same points as $\overrightarrow{AC}$.
    c. There are an infinite number of points on $\overleftrightarrow{AB}$.
    d. choices b and c only
    e. choices a and c only

14. Given the following figure, which of the statements is true?

    a. $\overline{XY}$ has a definite length.
    b. There are exactly two points on $\overline{XY}$.
    c. $\overline{XY}$ is the same as $\overleftrightarrow{XY}$.
    d. choices a and c only
    e. choices a and b only

15. Given the following figure, which of the statements is true?

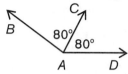

    a. The measure of $\angle CAD$ is twice the measure of $\angle BAD$.
    b. $\angle BAC \cong \angle CAD$
    c. The measure of $\angle BAD$ is one half the measure of $\angle BAC$.
    d. All of the above are true.
    e. choices **a** and **b** only

16. Given that M is the midpoint of $\overline{CD}$, the length of $\overline{CM} = 3x + 7$, and the length of $\overline{CD} = 32$ cm, what is the value of x?
    a. 3 cm
    b. $8\frac{1}{2}$ cm
    c. $4\frac{1}{6}$ cm
    d. $7\frac{2}{3}$ cm
    e. 6 cm

17. $\overrightarrow{RS}$ bisects $\angle QRT$, the measure of $\angle QRS = 5x + 35$, and the measure of $\angle TRS = 10x + 15$. What is the measure of $\angle QRT$?
    a. 55°
    b. 4°
    c. 110°
    d. $(5x + 20)°$
    e. $(5x - 20)°$

18. $\overline{ST}$ bisects $\overline{PQ}$, they intersect at point R, and the measure of $\overline{ST} = 12$ mm. If $\overline{PR} = 6x$ and $\overline{RQ} = 4x + 4$, what is the value of x?
    a. 3.2 mm
    b. 1 mm
    c. 2 mm
    d. 4.2 mm
    e. 18 mm

19. In the figure below, if $\overline{AC}$ = 31 cm, what is the value of x?

A   B   C

    $2x + 1$ $3x$

    a. $\frac{29}{4}$ cm

    b. 6 cm

    c. 3 cm

    d. 15.5 cm

    e. $\frac{31}{3}$ cm

20. If the measure of $\angle LMN = 85°$, and $\overrightarrow{MQ}$ bisects $\angle LMN$, which of the following is true?
    a. measure of $\angle LMQ = 85°$
    b. measure of $\angle QMN = 85°$
    c. measure of $\angle LMQ = 42.5°$
    d. measure of $\angle LMQ = 170°$
    e. None of the above statements is true.

21. $\overleftrightarrow{CD}$ bisects $\overline{AB}$ and they meet at point E. The measure of $\overline{AE} = 6x + 3$ inches, and the measure of $\overline{EB} = 4x + 17$ inches. What is the length of $\overline{AB}$?
    a. 7 inches
    b. 45 inches
    c. 90 inches
    d. 22.5 inches
    e. 14 inches

22. If the measure of $\overline{CD}$ = 7 inches and point A is the midpoint of $\overline{CD}$, what is the measure of $\overline{CA}$?
    a. 7 inches
    b. 14 inches
    c. 3.5 inches
    d. 42 inches
    e. 21 inches

23. $\angle FGJ$ is bisected by $\overrightarrow{GH}$, and the measure of $\angle FGJ = 150°$. If the measure of $\angle FGH = (2x + 8)°$, what is the measure of $\angle HGJ$?
    a. 33.5°
    b. 75°
    c. 79°
    d. 67°
    e. 35.5°

24. In the following figure, the measure of $\overline{WX} = 3x + 7$ meters, and the measure of $\overline{XY} = 5x - 5$ meters. Point X is the midpoint of $\overline{WY}$. What is the length of $\overline{WY}$?

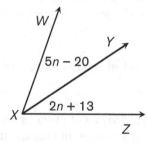

a. 50 m
b. $\frac{17}{7}$ m
c. 25 m
d. 12.5 m
e. $\frac{19}{6}$ m

25. In the following figure, $\overrightarrow{XY}$ bisects $\angle WXZ$. What is the measure of $\angle WXY$?

a. 11°
b. 35°
c. 55°
d. 70°
e. 17.5°

ANSWERS

Following are the answers and explanations to the practice quiz. Read them over carefully for any problems that you answered incorrectly.

1. c. Points R, W, and Z are not collinear. Statement **c** is false. All of the other statements about the figure are true.

2. d. Only statements **a** and **c** are true. Choice **a** is true—points W, X, and S are coplanar, since any three points define a plane. Choice **c** is true—the two planes listed are parallel, as they are the front and back sides of the rectangular solid. Choice **b** is false—when two planes intersect, they intersect in a line, not in a point.

3. b. $\overleftrightarrow{WX}$ and $\overleftrightarrow{SY}$ are skew lines. They are two lines that neither intersect nor are parallel. They do not lie in the same plane.

4. e. All of the choices listed are parallel line pairs. In all cases, they are a pair of lines in the same plane that do not intersect.

5. a. $\angle WZT$ measures 90°, as indicated by the small box in the interior of the angle. Note that choice **b** is incorrect. Even though the angle appears to be acute, the small box in the interior of the angle indicates that it is a right angle.

6. d. Both choice **a** and choice **c** show sets of coplanar points. Choice **c** is the plane that is a side of the solid. Choice **a** is a plane that runs diagonally through the solid.

7. a. The only statement that is true is that the named segment and line are perpendicular. $\overleftrightarrow{AB} \perp \overline{BY}$, as shown in the figure by the small box in the interior of $\angle YBA$.

8. b. The angle measures less than 180° and measures more than 90°. This classification is obtuse.

9. d. The only statement that is true is that $\angle ABC$ is congruent to $\angle DEF$. Choice **a** is incorrect, because a ray, not an angle, bisects an angle.

10. d. The three statements are all true given the facts, and the figure drawn.

11. e. None of the statements is true. In order to evaluate these statements, draw the perpendicular rays with endpoint C. This is the vertex of the angle.

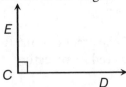

From the figure, it is clear that none of the three statements is true. Note that $\angle CED$ is an acute angle, since the vertex is E, not C. This can be seen if you draw in the side of the angle on the figure:

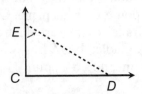

12. e. $\overrightarrow{TS}$ and $\overrightarrow{TR}$ name the exact same set of points since they are just two names for the same ray. They both have an infinite number of points, with endpoint T and in the direction of both S and R.

13. d. Statements **b** and **c** are both true. Statement **a** is false—you cannot measure the length of a line, it has an infinite length. Statement **b** is true—these are two names for the same line. Statement **c** is true—all lines have an infinite number of points.

14. a. All segments have a definite length, even though they contain an infinite number of points.

15. b. The two angles are congruent since they have the same measure.

16. a. In order to solve this problem, it is helpful to draw the segment with the given information included.

$$\overset{3x+7}{\underset{C \qquad M \qquad D}{\bullet\rule{3cm}{0.4pt}\bullet\rule{3cm}{0.4pt}\bullet}}$$

Since M is the midpoint of the segment, each half of the segment is represented by $3x + 7$, for a total length of 32 cm.

$3x + 7 + 3x + 7 = 32$	Set up the equation.
$6x + 14 = 32$	Combine like terms.
$6x + 14 - 14 = 32 - 14$	Subtract 14 from both sides.
$\frac{6x}{6} = \frac{18}{6}$	Combine like terms, and divide both sides by 6.
$x = 3$ cm	

17. c. Draw a picture to help visualize the problem.

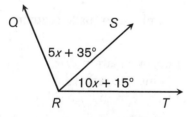

Since it is stated that $\overrightarrow{RS}$ bisects $\angle QRT$, $\angle QRS$ and $\angle TRS$ are congruent, and so have the same measure. First, find the value of x, and then use that value to calculate the value of the angle.

$5x + 35 = 10x + 15$	Set up an equation.
$5x - 5x + 35 = 10x - 5x + 15$	Subtract $5x$ from both sides.
$35 = 5x + 15$	Combine like terms.
$35 - 15 = 5x + 15 - 15$	Subtract 15 from both sides.
$\frac{20}{5} = \frac{5x}{5}$	Combine like terms, and divide both sides by 5.

$4 = x$

Using the value of 4 for x, substitute in to find the measure of $\angle QRS$: $5(4) + 35 = 20 + 35 = 55°$. This is one half of the measure of $\angle QRT$. The measure of $\angle QRT = 2(55) = 110°$.

18. c. Draw the figure to help solve the problem.

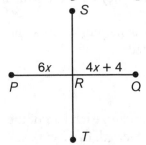

The fact that $\overline{ST} = 12$ mm does not help with the solution. Use algebra.

$6x = 4x + 4$	Set up an equation.
$6x - 4x = 4x - 4x + 4$	Subtract $4x$ from both sides.
$\frac{2x}{2} = \frac{4}{2}$	Combine like terms, and divide both sides by 2.

$x = 2$

19. b. The two segments together combine to make segment $\overline{AC} = 31$ cm. Use algebra.

$2x + 1 + 3x = 31$	Set up an equation.
$5x + 1 = 31$	Combine like terms.
$5x + 1 - 1 = 31 - 1$	Subtract 1 from both sides.
$\frac{5x}{5} = \frac{30}{5}$	Combine like terms, and divide both sides by 5.

$x = 6$ cm

20. c. Draw the angle with bisector. $\angle LMQ$ is half the measure of $\angle LMN$.

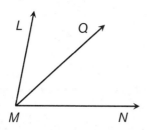

21. c. Draw the figure to help with the solution.

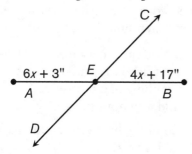

$\overline{AE} \cong \overline{EB}$, since E is the midpoint. Use algebra to solve for x, and then find the length of $\overline{AB}$.

$6x + 3 = 4x + 17$	Set up the equation.
$6x - 4x + 3 = 4x - 4x + 17$	Subtract $4x$ from both sides.
$2x + 3 = 17$	Combine like terms.
$2x + 3 - 3 = 17 - 3$	Subtract 3 from both sides.
$\frac{2x}{2} = \frac{14}{2}$	Combine like terms, and divide both sides by 2.
$x = 7$	

Now, substitute $x = 7$ to find the length of $\overline{AB} = (6 \times 7 + 3) + (4 \times 7 + 17) = 45 + 45 = 90$ inches.

22. c. If A is the midpoint of $\overline{CD}$, then the measure of $\overline{CA}$ is half the measure of $\overline{CD}$.

23. b. Make a sketch of the figure to help solve the problem.

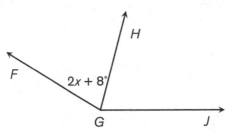

∠*HGJ* is half the measure of ∠*FGJ*, since it is bisected by $\overrightarrow{GH}$. The fact that ∠*FGH* = (2*x* + 8)° is not needed to solve the problem.

24. a. Since *X* is the midpoint, the two segments have the same length. Use algebra to solve for *x*, and then find the length of $\overline{WY}$.

$3x + 7 = 5x - 5$	Set up the equation.
$3x - 3x + 7 = 5x - 3x - 5$	Subtract $3x$ from both sides.
$7 = 2x - 5$	Combine like terms.
$7 + 5 = 2x - 5 + 5$	Add 5 to both sides.
$\frac{12}{2} = \frac{2x}{2}$	Combine like terms, and divide both sides by 2.
$6 = x$	

Now, $\overline{WY} = 3x + 7 + 5x - 5 = 8x + 2$, by combining like terms. Substitute in 6 for *x*, to get the length: $8 \times 6 + 2 = 48 + 2 = 50$ meters.

25. b. ∠*WXY* ≅ ∠*YXZ*, so the measures are equal. Use algebra to solve for *n*, and then calculate the measure of ∠*WXY*.

$2n + 13 = 5n - 20$	Set up the equation.
$2n - 2n + 13 = 5n - 2n - 20$	Subtract $2n$ from both sides.
$13 = 3n - 20$	Combine like terms.
$13 + 20 = 3n - 20 + 20$	Add 20 to both sides.
$\frac{33}{3} = \frac{3n}{3}$	Combine like terms, and divide both sides by 3.
$11 = n$	

Now, ∠*WXY* = 5*n* − 20, so substitute in 11 for *n* to get:
$5 \times 11 - 20 = 55 - 20 = 35°$.

Special Angle Pairs and Angle Measurement

Frequently, on math tests that assess geometry skills, there are several special pairs of angles that occur. See how many of these pairs you remember by taking this ten-question benchmark quiz. Use this quiz to determine how much of your study time should be devoted to this chapter.

BENCHMARK QUIZ

1. In the following figure, what is the value of x?

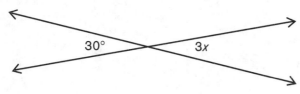

a. $0°$
b. $10°$
c. $90°$
d. $50°$
e. $30°$

2. The following diagram shows parallel lines cut by a transversal. What is the measure of ∠5?

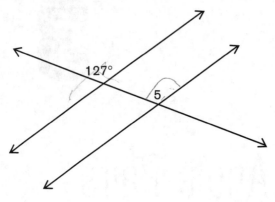

a. 100°
b. 53°
c. 127°
d. 37°
e. 90°

3. Which angle(s) shown are supplementary to ∠2, in the following parallel lines cut by a transversal?

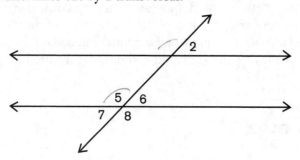

a. ∠6
b. ∠6 and ∠7
c. ∠5
d. ∠8
e. ∠5 and ∠8

4. The following figure shows parallel lines cut by a transversal. What is the measure of $\angle 8$?

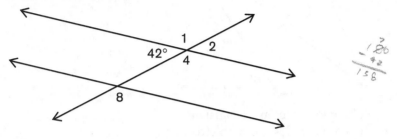

a. 48°
b. 138°
c. 42°
d. 148°
e. 118°

5. What are the values of the two angles?

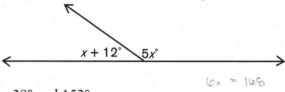

a. 28° and 152°
b. 40° and 140°
c. 28° and 62°
d. 40° and 50°
e. 45° and 135°

6. What is the measure of $\angle ADB$?

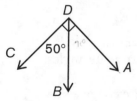

a. 90°
b. 50°
c. 40°
d. 130°
e. 45°

7. Which of the statements below is true?

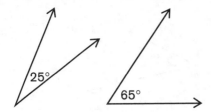

    a. The two angles are congruent.
    b. The two angles are supplementary.
    c. The two angles are vertical.
    d. The two angles are complementary.
    e. The two angles are obtuse.

8. What is the value of x, given that the lines are parallel lines cut by a transversal?

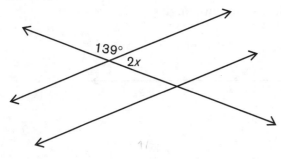

    a. 20.5°
    b. 69.5°
    c. 41°
    d. 139°
    e. 180°

9. Which statement below is ALWAYS true?
    a. Both angles of a supplementary pair are acute angles.
    b. The angles in a vertical pair are acute.
    c. The angles in a complementary pair are congruent.
    d. One of the angles in a supplementary pair is obtuse.
    e. The angles in a vertical pair are congruent.

10. Using the figure of parallel lines cut by a transversal below, which of the following statements is false?

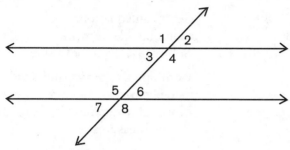

   a. ∠8 and ∠2 are supplementary.
   b. ∠1 and ∠4 are vertical.
   c. ∠2 and ∠7 are congruent.
   d. ∠5 and ∠1 are supplementary.
   e. ∠1 and ∠4 are congruent.

BENCHMARK QUIZ SOLUTIONS

Use the answer key to check your knowledge of special angle pairs. Then, read the suggestions following the answer explanations to plan your study of this chapter.

1. b. The two angles marked are a vertical pair, and vertical angles are congruent. Use algebra:

$$3x = 30 \qquad \text{Set up an equation.}$$
$$\frac{3x}{3} = \frac{30}{3} \qquad \text{Divide both sides by 3.}$$
$$x = 10$$

2. c. The two marked angles are corresponding angles, and corresponding angles are congruent; they have the same measure.

3. e. ∠2 is an acute angle, thus any angle that is obtuse will be supplementary.

4. b. ∠8 and ∠1 are alternate exterior angles, and are therefore congruent. Because ∠1 forms a linear pair with the angle of 42°, the measure of ∠1 = 180 − 42 = 138°. The measure of ∠8 = 138°.

5. b. The sum of the measure of these two angles is 180° due to the straight line they form. Use algebra:

$5x + x + 12 = 180$	Set up an equation.
$6x + 12 = 180$	Combine like terms.
$6x + 12 - 12 = 180 - 12$	Subtract 12 from both sides.
$\frac{6x}{6} = \frac{168}{6}$	Combine like terms, and divide each side by 6.
$x = 28$	Now, substitute in 28 for x to find the angle measures.

One angle is $5(28) = 140°$; the other angle is $28 + 12 = 40°$.

6. c. The two angles shown are adjacent angles that form a right angle. The measure of $\angle ADB = 90 - 50 = 40°$.

7. d. Any two angles whose sum of degree-measure equals 90° are complementary.

8. a. The two marked angles are supplementary. Use algebra:

$2x + 139 = 180$	Set up the equation.
$2x + 139 - 139 = 180 - 139$	Subtract 139 from both sides.
$\frac{2x}{2} = \frac{41}{2}$	Combine like terms, and divide both sides by 2.
$x = 20.5$	

9. e. Vertical angles are always congruent. Note that choice **d** is not always true; the two angles can both measure 90°.

10. d. $\angle 1$ and $\angle 5$ are not supplementary. These angles are congruent, as they are corresponding angles.

BENCHMARK QUIZ RESULTS

If you answered 8–10 questions correctly, you have a solid understanding of angle pairs; you may not need to concentrate your efforts on this topic. Skim over the chapter, but take your time on the sections in which you need to be refreshed. It is suggested that you take the quiz at the end of this chapter to make sure that you can handle different problem situations.

If you answered 4–7 questions correctly, you need a more careful review of these angle pairs. Carefully read through the lessons in this chapter for review and skill building. Pay attention to the sidebars that refer you to

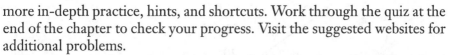
more in-depth practice, hints, and shortcuts. Work through the quiz at the end of the chapter to check your progress. Visit the suggested websites for additional problems.

If you answered 1–3 questions correctly, you need to take time to read through the entire lesson in this chapter. Concentrate on the sidebars and visuals that will aid comprehension. Go to the suggested websites in the Extra Help sidebar in this chapter and do extended practice. You may also want to refer to *Geometry Success in 20 Minutes a Day*, Lessons 4 and 5, published by LearningExpress.

JUST IN TIME LESSON—SPECIAL ANGLE PAIRS AND ANGLE MEASUREMENT

Topics in this chapter include:

- Complementary and Supplementary Angles
- Vertical Angles
- Angle Pairs Formed by Parallel Lines Cut by a Transversal
- Solving Geometry Problems by Using Algebra

Angles were introduced in Chapter 2. In this chapter, the focus will be on several pairs of special angle relationships that play an important role in the development of geometry.

COMPLEMENTARY AND SUPPLEMENTARY ANGLES

GLOSSARY
COMPLEMENTARY ANGLES two angles in which the sum of their measures equals 90°
SUPPLEMENTARY ANGLES two angles in which the sum of their measures equals 180°

Complementary and supplementary angles may or may not be next to each other. The basis of classification for these angles is the sum of their measures.

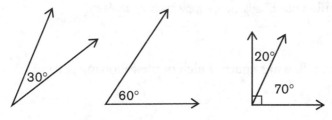

Complementary Angles

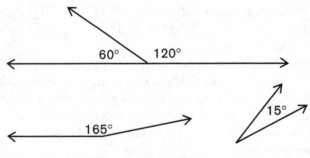

Supplementary Angles

Examples:

What is the measure of the complement of 23°?
Because an angle and its complement add up to 90°, the measure of the complement of 23° is 90 – 23 = 67°.

What is the measure of the supplement of 102°?
Because an angle and its supplement add up to 180°, the measure of the supplement of 102° is 180 – 102 = 78°.

GLOSSARY

ADJACENT ANGLES angles that share a common vertex, a common side and share no interior points

Example:

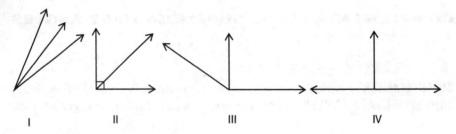

LINEAR PAIR a pair of adjacent, supplementary angles

Example:

In the following figure, which of the following is true?

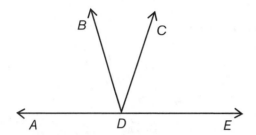

a. ∠CDB and ∠CDE are supplementary.
b. ∠CDB and ∠ADB are a linear pair.
c. ∠ADB and ∠EDB are a linear pair.
d. The sum of the measures of ∠ADB and ∠EDB is 100°.
e. None of the above is true.

Choice **c** is the correct answer. These two angles are supplementary and adjacent. Choice **a** is false because the two angles are adjacent but together do not form a straight line. Choice **b** is false because the two angles are adjacent but are not supplementary. Choice **d** is false because the sum of these angles is equal to 180°.

SHORTCUT

Any two adjacent angles that form a straight line are supplementary. Any two adjacent angles that form a right angle are complementary. In the figures above, Figure II shows complementary adjacent angles, and Figure IV shows supplementary adjacent angles.

VERTICAL ANGLES

When two distinct lines intersect, four angles are formed: ∠1, ∠2, ∠3, and ∠4.

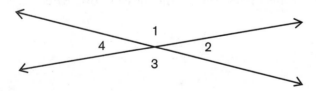

 GLOSSARY

VERTICAL ANGLES two non-adjacent angles that are formed by the intersection of two lines, rays, or segments

In the figure above, the pairs of non-adjacent angles (one pair is angles 1 and 3; the other pair is angles 2 and 4) are called *vertical angles*.

RULE BOOK

Vertical angles are congruent.

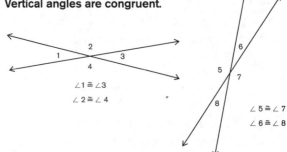

$\angle 1 \cong \angle 3$
$\angle 2 \cong \angle 4$

$\angle 5 \cong \angle 7$
$\angle 6 \cong \angle 8$

Example:
In the following diagram, what is the measure of $\angle 2$?

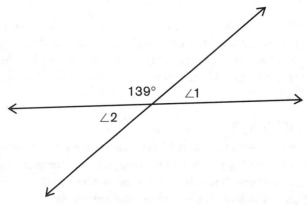

$\angle 1$ and the angle marked 139° are supplementary. Their measure adds to 180°. So the measure of $\angle 1 = 180 - 139 = 41°$. Because $\angle 2$ is a vertical pair with $\angle 1$, it is also equal to 41°.

Example:
Given the following angles, what is the measure of $\angle ABE$?

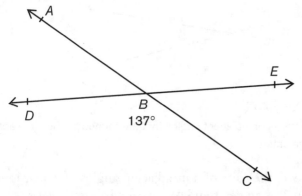

$\angle ABE$ is a vertical pair with $\angle DBC$, whose measure is 137°. So, the measure of $\angle ABE = 137°$.

ANGLE PAIRS FORMED BY PARALLEL LINES CUT BY A TRANSVERSAL

When two parallel lines are given in a figure, there are two main areas: the interior and the exterior.

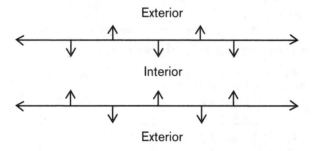

When two parallel lines are cut by a third line, the third line is called the *transversal*. In the example below, eight angles are formed when parallel lines *m* and *n* are cut by a transversal line, *t*.

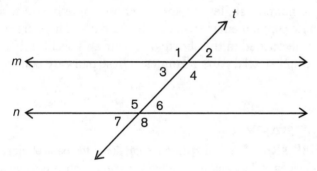

There are several special pairs of angles formed from this figure. Some pairs have already been reviewed:

Vertical pairs: ∠1 and ∠4
                      ∠2 and ∠3
                      ∠5 and ∠8
                      ∠6 and ∠7

Recall that all pairs of vertical angles are congruent.

Supplementary pairs: ∠1 and ∠2
                          ∠2 and ∠4
                          ∠3 and ∠4
                          ∠1 and ∠3
                          ∠5 and ∠6
                          ∠6 and ∠8
                          ∠7 and ∠8
                          ∠5 and ∠7

Recall that supplementary angles are angles whose angle measure adds up to 180°. All of these supplementary pairs are linear pairs. There are other supplementary pairs described in the shortcut later in this section.

There are three other special pairs of angles. These pairs are congruent pairs.

●●●●●●● GLOSSARY

ALTERNATE INTERIOR ANGLES   two angles in the interior of the parallel lines, and on opposite (alternate) sides of the transversal. Alternate interior angles are non-adjacent and congruent.

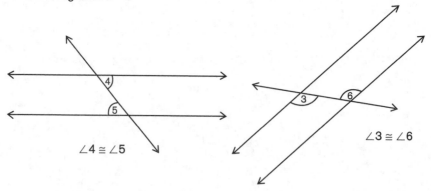

∠4 ≅ ∠5   ∠3 ≅ ∠6

ALTERNATE EXTERIOR ANGLES two angles in the exterior of the parallel lines, and on opposite (alternate) sides of the transversal. Alternate exterior angles are non-adjacent and congruent.

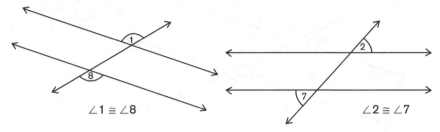

CORRESPONDING ANGLES two angles, one in the interior and one in the exterior, that are on the same side of the transversal. Corresponding angles are non-adjacent and congruent.

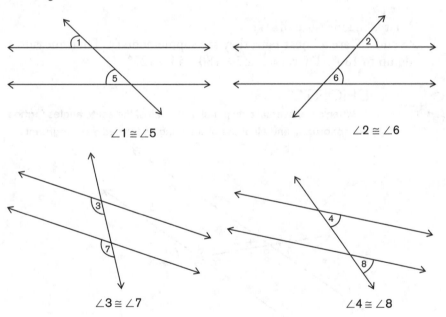

Use the following diagram of parallel lines cut by a transversal to answer the example problems.

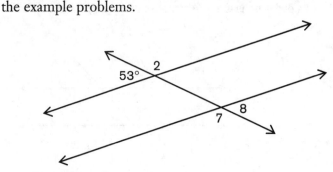

Example:
What is the measure of ∠8?
The angle marked with measure 53° and ∠8 are alternate exterior angles. They are in the exterior, on opposite sides of the transversal. Because they are congruent, the measure of ∠8 = 53°.

Example:
What is the measure of ∠7?
∠8 and ∠7 are a linear pair; they are supplementary. Their measures add up to 180°. Therefore, ∠7 = 180 – 53 = 127°.

SHORTCUTS

1. When a transversal cuts parallel lines, all of the acute angles formed are congruent, and all of the obtuse angles formed are congruent.

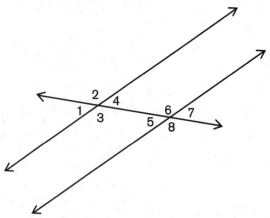

In the figure above, ∠1, ∠4, ∠5, and ∠7 are all acute angles. They are all congruent to each other. ∠1 ≅ ∠4 are vertical angles. ∠4 ≅ ∠5 are alternate interior angles, and ∠5 ≅ ∠7 are vertical angles. The same reasoning applies to the obtuse angles in the figure: ∠2, ∠3, ∠6, and ∠8 are all congruent to each other.

2. When parallel lines are cut by a transversal line, any one acute angle formed and any one obtuse angle formed are supplementary.

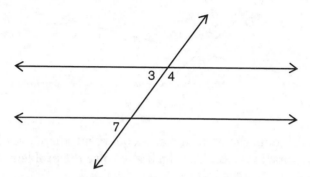

From the figure, you can see that ∠3 and ∠4 are supplementary because they are a linear pair. Notice also that ∠3 ≅ ∠7, since they are corresponding angles. Therefore, you can substitute ∠7 for ∠3 and know that ∠7 and ∠4 are supplementary.

Example:
In the following figure, there are two parallel lines cut by a transversal. Which marked angle is supplementary to ∠1?

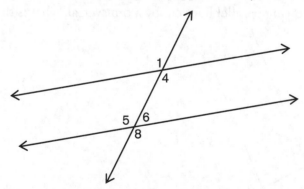

The angle supplementary to ∠1 is ∠6. ∠1 is an obtuse angle, and any one acute angle, paired with any obtuse angle are supplementary angles. This is the only angle marked that is acute.

SOLVING GEOMETRY PROBLEMS BY USING ALGEBRA

Often, problems will test your knowledge of these angle pairs and your skill in equation solving.

Example:
Use the following diagram to give the measure of ∠*DEC*.

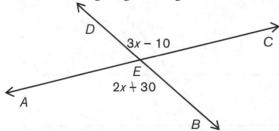

In the diagram, the vertical pair is congruent; therefore each angle has the same measure. Use algebra to solve the problem:

$2x + 30 = 3x - 10$	Set up the algebraic equation.
$2x - 2x + 30 = 3x - 2x - 10$	Subtract $2x$ from both sides.
$30 = x - 10$	Combine like terms.
$30 + 10 = x - 10 + 10$	Add 10 to both sides.
$40 = x$	

Now, substitute 40 into the original expression for ∠*DEC*:
∠*DEC* = 3(40) − 10 = 120 − 10 = 110°.

Example:
Given the following parallel lines cut by a transversal, what is the value of *x*?

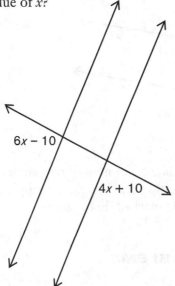

In this diagram, the two marked angles appear to be almost congruent. It is difficult to establish if they are congruent or not. However, they are both exterior angles on the same side of the transversal. They are NOT congruent. The two marked angles are supplementary. Use algebra:

$(6x - 10) + (4x + 10) = 180$ Set up an equation to find x.

$\dfrac{10x}{10} = \dfrac{180}{10}$ Combine like terms, and divide both sides by 10.

$x = 18$

Example:
Use the figure below of parallel lines cut by a transversal to find the measure of $\angle 2$.

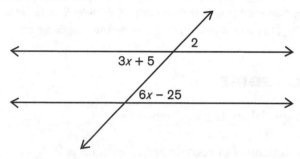

Because all three angles marked are acute angles, they are all congruent. Use algebra:

$3x + 5 = 6x - 25$	Set up an equation.
$3x + 5 + 25 = 6x - 25 + 25$	Add 25 to both sides.
$3x + 30 = 6x$	Combine like terms.
$3x - 3x + 30 = 6x - 3x$	Subtract $3x$ from both sides.
$\dfrac{30}{3} = \dfrac{3x}{3}$	Combine like terms, and divide both sides by 3.
$10 = x$	

Now, use this value to find the measure of the angle. $\angle 2 = 3x + 5 = (3 \times 10) + 5 = 35°$.

When using algebra to solve problems, be sure what the problem is asking for. Sometimes, the value of the variable is sought. Other times, the measure of the angle is required.

EXTRA HELP

To practice identifying the angles formed by parallel lines and a transversal, visit the website *http://studyworksonline.com*. Click on *Explorations* at the top of the website. Then, click on *Interactive Geometry* on the left hand side. Finally, click on *Alternate Angles*. Another helpful site is *www.shodor.org/interactivate/activities/angles/index.html#*. You will see an activity that tests your knowledge of angle classification as well as the identification of the angle types. Another useful site is *www.math.com*. On the left sidebar, click on *Geometry.* Then, under *Basic Building Blocks of Geometry* click on *Classifying Angles.* There is a series of interactive lessons, followed by a short interactive quiz. You may also want to reference the book *Geometry Success in 20 Minutes a Day,* Lessons 4 and 5, published by LearningExpress.

TIPS AND STRATEGIES

When working with special angle pairs, remember:

- The sum of the measures of complementary angles is 90°.
- The sum of the measures of supplementary angles is 180°.
- Two adjacent supplementary angles form a linear pair.
- Vertical angles are angles across from each other, when formed by intersecting lines, rays, or segments.
- Vertical angles are congruent.
- There are special angle pairs formed by two parallel lines cut by a transversal:

    Alternate interior angles are congruent.

    Alternate exterior angles are congruent.

    Corresponding angles are congruent.
- All of the acute angles formed by two parallel lines cut by a transversal are congruent.
- All of the obtuse angles formed by two parallel lines cut by a transversal are congruent.
- When parallel lines are cut by a transversal, any one acute angle and any one obtuse angle are supplementary.

PRACTICE QUIZ

Try these 25 problems. When you are finished, review the answers and explanations to see if you have mastered this concept.

1. In the following figure of two intersecting lines, what is the measure of ∠2?

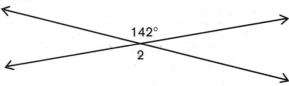

    a. 38°
    b. 58°
    c. 71°
    d. 142°
    e. 42°

2. What is the value of *x*, shown below?

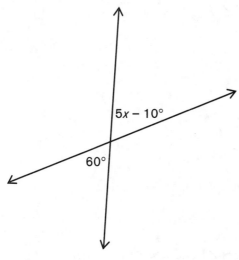

    a. 60°
    b. 26°
    c. 14°
    d. 10°
    e. 120°

JUST IN TIME GEOMETRY

Use the following diagram of parallel lines cut by a transversal to answer questions 3 through 9.

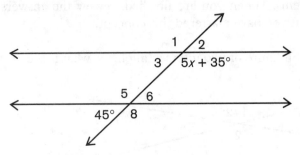

3. What is the measure of ∠8?
   a. 45°
   b. 145°
   c. 55°
   d. 135°
   e. 155°

4. What is the measure of ∠3?
   a. 45°
   b. 145°
   c. 135°
   d. 55°
   e. 155°

5. What is the measure of ∠6?
   a. 45°
   b. 135°
   c. 180°
   d. 55°
   e. 100°

6. What type of angle pair is ∠1 and ∠5?
   a. alternate interior angles
   b. supplementary angles
   c. corresponding angles
   d. alternate exterior angles
   e. vertical angles

7. What is the measure of $\angle 2$?
  a. 55°
  b. 180°
  c. 80°
  d. 45°
  e. 135°

8. What type of angle pair is $\angle 1$ and $\angle 3$?
  a. alternate interior angles
  b. supplementary angles
  c. corresponding angles
  d. alternate exterior angles
  e. vertical angles

9. In the diagram, what is the value of x?
  a. 21°
  b. 27°
  c. 4°
  d. 135°
  e. 20°

10. What are the values of the two angles shown in the figure below?

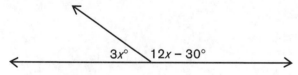

  a. 18° and 152°
  b. 30° and 150°
  c. 45° and 135°
  d. 42° and 138°
  e. 60° and 120°

11. What is the measure of the complement of 62°?
  a. 38°
  b. 28°
  c. 118°
  d. 128°
  e. 62°

12. Which of the following statements is ALWAYS TRUE when parallel lines are cut by a transversal?
　a. The sum of the degree measure of corresponding angles is 180°.
　b. The sum of the degree measure of complementary angles is 180°.
　c. The angles in a vertical pair are acute.
　d. Corresponding angles are supplementary.
　e. Corresponding angles are congruent.

13. In the figure below of parallel lines cut by a transversal, which angle is supplementary to ∠8?

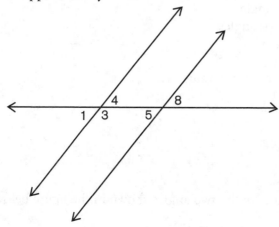

　a. ∠1
　b. ∠3
　c. ∠4
　d. ∠5
　e. none of the above

14. Using the figure below, which of the statements is true?

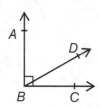

　a. $\angle ABD \cong \angle DBC$
　b. $\angle ABD$ and $\angle ABC$ are adjacent angles.
　c. $\angle ABD$ and $\angle DBC$ are complementary.
　d. $\angle DBC$ is half of the measure of $\angle ABC$.
　e. All of the above are true.

15. What is the value of x?

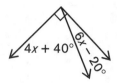

a. 7°
b. 10°
c. 9°
d. 68°
e. 22°

16. Use the following figure to determine which statements is true.

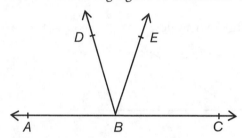

a. $\angle ABD$ and $\angle DBC$ are a linear pair. –
b. $\angle ABE$ and $\angle EBC$ are adjacent. –
c. $\angle DBE$ and $\angle EBC$ are adjacent. –
d. $\angle EBC$ and $\angle EBA$ are a linear pair. –
e. All of the above are true.

17. What is the measure of $\angle ABC$?

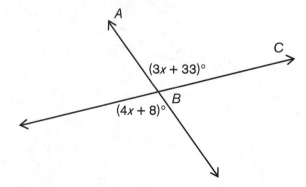

a. 25°
b. 75°
c. 108°
d. 150°
e. 100°

18. Which of the following statements is true about the figure?

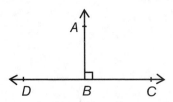

    a. ∠*DBC* measures 180°.
    b. ∠*BAC* is a right angle.
    c. ∠*CBA* and ∠*ABD* are supplementary.
    d. choices **a** and **c**
    e. All of the above are true.

19. What are the values of the two angles?

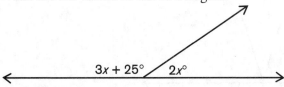

    a. 31° and 159°
    b. 62° and 118°
    c. 28° and 62°
    d. 31° and 149°
    e. 31° and 69°

20. What is the value of *x* in the figure below?

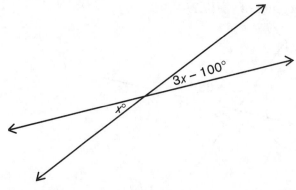

    a. 150°
    b. 50°
    c. 100°
    d. 130°
    e. 25°

21. In the diagram of parallel lines cut by a transversal, shown below, which of the following statements is false?

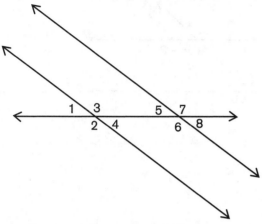

   a. ∠3 and ∠4 are vertical angles.
   b. ∠5 and ∠8 are corresponding angles.
   c. ∠3 and ∠5 are alternate interior angles.
   d. ∠2 and ∠8 are alternate exterior angles.
   e. All of the above are false.

22. The following diagram shows parallel lines cut by a transversal. What is the measure of ∠2?

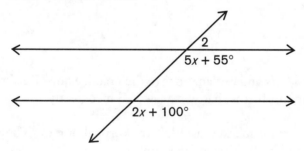

   a. 80°
   b. 100°
   c. 20°
   d. 160°
   e. 50°

23. Which of the statements below is ALWAYS TRUE?
   a. Supplementary angles are both obtuse angles.
   b. Complementary angles are both acute angles.
   c. At least one of the angles in a supplementary pair is obtuse.
   d. Alternate exterior angles are both obtuse angles.
   e. Any two acute angles are complementary.

24. The following diagram shows parallel lines cut by a transversal. What is the value of x?

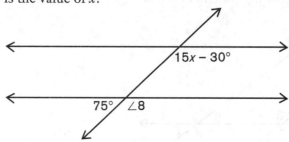

$15x - 30°$

$75°$ $\angle 8$

   a. $9°$
   b. $75°$
   c. $115°$
   d. $3°$
   e. $5°$

25. Two angles are a linear pair. Their measures are represented by $x + 10$ and $3x + 10$. What are the measures of the angles?
   a. $40°$ and $40°$
   b. $40°$ and $140°$
   c. $50°$ and $130°$
   d. $60°$ and $120°$
   e. $90°$ and $90°$

ANSWERS

Following are the answers and explanations to the practice quiz. Read them over carefully for any problems that you answered incorrectly.

   1. d. $\angle 2$ and the angle marked as $142°$ are vertical angles. They are congruent and have the same measure.

   2. c. This is a pair of vertical angles. Their angle measures are equal. Use algebra to find the value of x:

$5x - 10 = 60$	Set up the equation.
$5x - 10 + 10 = 60 + 10$	Add 10 to both sides.
$\dfrac{5x}{5} = \dfrac{70}{5}$	Combine like terms, and divide both sides by 5.
$x = 14$	

3. d. The angle marked with the 45° measure and ∠8 are supplementary. The sum of their angle measure is 180°.

$45 + m\angle8 = 180$ Set up the equation.
$45 + m\angle8 - 45 = 180 - 45$ Subtract 45 from both sides.
$m\angle8 = 135°$

4. a. The corresponding angles are the angle marked with the 45° measure and ∠3. They are congruent; their angle measures are equal.

5. a. The angle marked with the 45° measure and ∠6 are vertical angles. They are congruent; their angle measures are equal.

6. c. ∠1 and ∠5 are corresponding angles. One is in the interior, one is in the exterior, and they are on the same side of the transversal.

7. d. The angle marked with the 45° measure and ∠2 are alternate exterior angles. They are congruent; their angle measures are equal.

8. b. ∠1 and ∠3 are supplementary. Together, they form a straight line, whose angle measure is 180°.

9. e. When two parallel lines are cut by a transversal, any one acute angle, together with any one obtuse angle, are supplementary. Use algebra:

$5x + 35 + 45 = 180$ Set up an equation.
$5x + 80 = 180$ Combine like terms.
$5x + 80 - 80 = 180 - 80$ Subtract 80 from both sides of the equation.
$\frac{5x}{5} = \frac{100}{5}$ Combine like terms, and divide both sides by 5.

$x = 20$

10. d. These angles form a linear pair. The sum of their degree measures equals 180°. Use algebra:

$3x + 12x - 30 = 180$	Set up an equation.
$15x - 30 = 180$	Combine like terms.
$15x - 30 + 30 = 180 + 30$	Add 30 to both sides.
$\dfrac{15x}{15} = \dfrac{210}{15}$	Combine like terms, and divide both sides by 15.
$x = 14$	

Now, use this value for x to find the angle measures. One angle is $3x = 3 \times 14 = 42°$. The other angle is $12x - 30 = (12 \times 14) - 30 = 168 - 30 = 138°$.

11. b. An angle and its complement have angle measures whose sum is 90°. Let x stand for the complement of 62°, and use algebra:

$x + 62 = 90$	Set up the equation.
$x + 62 - 62 = 90 - 62$	Subtract 62 from both sides.
$x = 28°$	

12. e. The only statement that is always true is choice **e**. Choice **a** is only true if the two angles have a 90° measure. Choice **b** is never true. Choices **c** and **d** are sometimes true. The only time that corresponding angles are supplementary is when they have the measure of 90° each.

13. b. The only angle listed that is supplementary to ∠8 is the only obtuse angle marked, ∠3.

14. c. The only statement that is true is that the sum of the angle measures of ∠ABD and ∠DBC is 90°. Therefore, they are complementary.

15. a. The two angles marked are complementary because the sum of their angle measures is 90°. Use algebra to find the value of x:

$4x + 40 + 6x - 20 = 90$	Set up the equation.
$10x + 20 = 90$	Combine like terms.
$10x + 20 - 20 = 90 - 20$	Subtract 20 from both sides.
$\dfrac{10x}{10} = \dfrac{70}{10}$	Combine like terms, and divide both sides by 10.
$x = 7$	

16. e. All of the statements are true. Adjacent angles share a common side, and no interior points. Linear pairs are two adjacent angles whose sum of degrees is 180°, which is a straight line.

17. c. The two marked angles in the figure are vertical angles. Use algebra to solve for the variable x, and then find the measure of the angle.

$4x + 8 = 3x + 33$	Set up an equation.
$4x - 3x + 8 = 3x - 3x + 33$	Subtract $3x$ from both sides.
$x + 8 = 33$	Combine like terms.
$x + 8 - 8 = 33 - 8$	Subtract 8 from both sides.
$x = 25$	

Use the value of 25 in the angle expression $3x + 33$ to find the measure.

$(3 \times 25) + 33 = 108°$

18. d. Choices **a** and **c** are true. $\angle DBC$ measures 180° because it forms a straight line. The two angles $\angle CBA$ and $\angle ABD$ are supplementary as they form $\angle DBC$. Note that $\angle BAC$ is NOT a right angle. Point B is the vertex of this angle, not point A.

19. b. The two angles are a linear pair, so their angle measures add up to 180°. Use algebra to solve for x, and then find the angle measures.

$3x + 25 + 2x = 180$	Set up the equation.
$5x + 25 = 180$	Combine like terms.
$5x + 25 - 25 = 180 - 25$	Subtract 25 from both sides of the equation.
$\frac{5x}{5} = \frac{155}{5}$	Combine like terms, and divide both sides by 5.
$x = 31$	

Substitute the value of 31 into the expressions to find the angle measures.

$2 \times 31 = 62°$, and $(3 \times 31) + 25 = 118°$

20. b. The two marked angles are vertical angles, and their measures are equal. Use algebra to solve for x.

$x = 3x - 100$ Set up an equation.
$x - 3x = 3x - 3x - 100$ Subtract $3x$ from both sides.
$-2x = -100$ Combine like terms.
$\frac{-2x}{-2} = \frac{-100}{-2}$ Divide both sides by -2.
$x = 50°$

21. e. All of the statements are false. $\angle 3$ and $\angle 4$ are supplementary angles. $\angle 5$ and $\angle 8$ are vertical angles. $\angle 3$ and $\angle 5$ are interior angles on the same side of the transversal. $\angle 2$ and $\angle 8$ are exterior angles on the same side of the transversal.

22. e. The two angles marked by algebraic expressions are corresponding angles and their measures are equal. $\angle 2$ and $5x + 55$ are supplementary. First, use algebra to solve for x and then find the measure of $5x + 55$. Finally, find the measure of $\angle 2$.

$5x + 55 = 2x + 100$ Set up an equation.
$5x - 2x + 55 = 2x - 2x + 100$ Subtract $2x$ from both sides.
$3x + 55 = 100$ Combine like terms.
$3x + 55 - 55 = 100 - 55$ Subtract 55 from both sides.
$3x = 45$ Combine like terms.
$\frac{3x}{3} = \frac{45}{3}$ Divide both sides by 3.
$x = 15°$

The angle marked as $5x + 55 = (5 \times 15) + 55 = 130°$.
$\angle 2 = 180 - 130 = 50°$

23. b. Choice **b** is always true. Complementary angles must each be less than 90°, since the sum of their measures must equal 90°. Supplementary angles do not have to be obtuse. Two angles of 90° are supplementary. Alternate exterior angles may be either obtuse or acute.

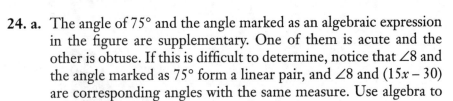

24. a. The angle of 75° and the angle marked as an algebraic expression in the figure are supplementary. One of them is acute and the other is obtuse. If this is difficult to determine, notice that ∠8 and the angle marked as 75° form a linear pair, and ∠8 and $(15x - 30)$ are corresponding angles with the same measure. Use algebra to solve for x.

$15x - 30 + 75 = 180$	Set up an equation.
$15x + 45 = 180$	Combine like terms.
$15x + 45 - 45 = 180 - 45$	Subtract 45 from both sides.
$15x = 135$	Combine like terms.
$\frac{15x}{15} = \frac{135}{15}$	Divide both sides by 15.
$x = 9°$	

25. c. Two angles that are a linear pair have angle measures that add up to 180°. Use algebra to solve for x, and then find the measure of the angles by substituting in the value for x.

$x + 10 + 3x + 10 = 180$	Set up an equation.
$4x + 20 = 180$	Combine like terms.
$4x + 20 - 20 = 180 - 20$	Subtract 20 from both sides.
$4x = 160$	Combine like terms.
$\frac{4x}{4} = \frac{160}{4}$	Divide both sides by 4.
$x = 40°$	

The angle $x + 10$ is $40 + 10 = 50°$, and the other angle $3x + 10$ is $(3 \times 40) + 10 = 120 + 10 = 130°$.

Triangles

The triangle is the fundamental figure in geometry. Virtually all tests that include geometry will have problems relating to triangles, so it's important to have a strong foundation in this area. Before studying the lessons in this chapter, take this ten-question benchmark quiz. When you are finished, check the answer key carefully. Follow the suggestions that follow which are based on how you scored on the quiz.

BENCHMARK QUIZ

1. Which of the following side measures could NOT form a triangle?
   a. 6, 8, 10
   b. 5, 5, 11
   c. 8, 12, 13
   d. 2, 7, 8
   e. 9, 9, 15

2. What is the classification of the following triangle?

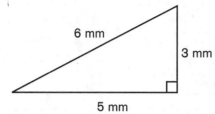

6 mm

3 mm

5 mm

    a. right isosceles
    b. acute scalene
    c. obtuse isosceles
    d. acute isosceles
    e. right scalene

3. Using the triangle below, what is the measure of ∠*XYZ*?

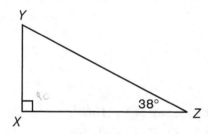

    a. 90°
    b. 52°
    c. 38°
    d. 142°
    e. 62°

4. By which method are the two triangles, Δ*WXZ* and Δ*YXZ* congruent?

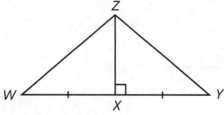

    a. SSS
    b. AAS
    c. SAS
    d. SSA
    e. cannot be determined

5. Given the following triangle, what is the value of x?

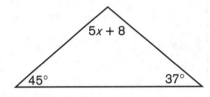

a. 90°
b. 14.8°
c. 18°
d. 2°
e. 54°

6. Given the triangle below, which of the following statements is true?

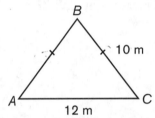

a. The measure of ∠*BAC* > the measure of ∠*ABC*.
b. The measure of ∠*ABC* > the measure of ∠*BCA*.
c. The measure of ∠*BAC* = the measure of ∠*BCA*.
d. choices a and c
e. choices b and c

7. Given △*XYZ* with exterior angle shown, what is the value of x?

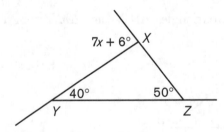

a. 12°
b. 170°
c. 90°
d. 25°
e. 20°

8. Given the following triangle, which of the following statements is FALSE?

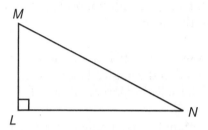

a. The altitude to $\overline{LN}$ is equal to the median of $\overline{LN}$.
b. $\overline{LN}$ is an altitude to the triangle.
c. The measure of $\angle LMN$ plus the measure of $\angle LNM$ is equal to 90°.
d. All of the above are true.
e. All of the above are false.

9. The angle measures in a triangle are in the ratio of 1 : 2 : 5. What is the measure of the largest angle?
a. 50°
b. 22.5°
c. 100°
d. 112.5°
e. 45°

$x + 2x + 5x = 180$

$6x = 180$

10. Use the following figure to find the value of x.

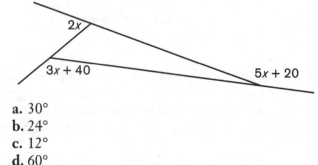

a. 30°
b. 24°
c. 12°
d. 60°
e. 130°

BENCHMARK QUIZ SOLUTIONS

Carefully check your answers, and read through the answer explanations. Grade yourself, and then follow the suggestions given under *Benchmark Quiz Results*.

1. b. The side measures of 5, 5, and 11 cannot form a triangle. If you add the congruent sides, $5 + 5 = 10$, this is not greater than the third side of 11. The sum of any two sides of a triangle must be greater than the measure of the third side.

2. e. The triangle is a right triangle, since it has an angle of 90°, as shown by the small box marked in the angle. The classification by sides is scalene, since all of the side measurements are different.

3. b. The acute angles of a right triangle are complementary. The sum of the measures of $\angle XYZ$ and 38° must equal 90°. The measure of $\angle XYZ = 90 - 38 = 52°$.

4. c. The two triangles both have an angle of 90°, which are congruent. $\overline{WX} \cong \overline{XY}$ as marked by the single slashes. The triangles share a common side of $\overline{ZX}$, so this is another congruent side. The angle is between the two congruent sides, so the correct method is side, angle, side or SAS.

5. c. The sum of the degree measures of the angles in a triangle equals 180°. Use algebra to solve for x.

$5x + 8 + 45 + 37 = 180$	Set up an equation.
$5x + 90 = 180$	Combine like terms.
$5x + 90 - 90 = 180 - 90$	Subtract 90 from both sides.
$5x = 90$	Combine like terms.
$\frac{5x}{5} = \frac{90}{5}$	Divide both sides by 5.
$x = 18°$	

6. e. Choices **b** and **c** are true. The largest angle is opposite to the largest side. Since $\overline{AC}$ is the largest side, $\angle ABC$ has the greatest measure. $\angle A$ and $\angle C$ are congruent because this is an isosceles triangle, and the base angles of an isosceles triangle are congruent.

7. a. The measure of an exterior angle is equal to the sum of the two remote interior angles. Use algebra to solve for x.

$7x + 6 = 40 + 50$	Set up an equation.
$7x + 6 = 90$	Combine like terms.
$7x + 6 - 6 = 90 - 6$	Subtract 6 from both sides.
$7x = 84$	Combine like terms.
$\frac{7x}{7} = \frac{84}{7}$	Divide both sides by 7.
$x = 12°$	

8. a. Choice **a** is the only false statement. $\overline{LM}$ is the altitude to side $\overline{LN}$ and the median (not shown in the figure) would extend from point M to the midpoint of $\overline{LN}$.

9. d. The angle measures are given as a ratio, so express them as x, $2x$, and $5x$, respectively. Together, their sum adds up to 180°. Use algebra to solve for x, and then substitute in this value for $5x$ to find the largest angle.

$x + 2x + 5x = 180$	Set up an equation.
$8x = 180$	Combine like terms.
$\dfrac{8x}{8} = \dfrac{180}{8}$	Divide both sides by 8.
$x = 22.5°$	

Use the value of 22.5 to find the largest angle, $5x$:
$5 \times 22.5 = 112.5°$.

10. a. The sum of the degree measures of the exterior angles of a triangle is 360°. Use algebra to solve for x.

$2x + 5x + 20 + 3x + 40 = 360$	Set up an equation.
$10x + 60 = 360$	Combine like terms.
$10x + 60 - 60 = 360 - 60$	Subtract 60 from both sides.
$10x = 300$	Combine like terms.
$\dfrac{10x}{10} = \dfrac{300}{10}$	Divide both sides by 10.
$x = 30°$	

BENCHMARK QUIZ RESULTS

If you answered 8–10 questions correctly, you have a good understanding of triangle relationships. Read over the chapter, concentrating on the areas where you were weak. Refer to the shortcut sidebars to pick up any techniques for saving time when solving triangle problems. Take the quiz at the end of this chapter to confirm your success with triangle relationship problems.

If you answered 4–7 questions correctly, you need to spend some time on this chapter. Read through the entire chapter; pay attention to the sidebars that refer you to more in depth practice, hints, and shortcuts. Work through the quiz at the end of the chapter to check your progress. The answer explanations will help you through any areas that still need clarification. Visit the suggested websites for additional practice.

If you answered 1–3 questions correctly, you need to devote your full attention to this chapter. Many problems in geometry deal with the triangle, so a complete understanding of triangles is crucial for any test that involves geometry. First, carefully read this chapter and concentrate on the sidebars and all figures. Be sure you know all the rules outlined in the Rule Books, and all of the necessary vocabulary. Go to the suggested websites in the Extra Help sidebar in this chapter, and do extended practice. For further study, refer to the book, *Geometry Success in 20 Minutes a Day*, Lessons 6 and 7, published by LearningExpress.

JUST IN TIME LESSON—TRIANGLES

Topics in this chapter include:

- Triangle Classification
- The Angles in a Triangle
- The Sides of a Triangle
- Median and Altitude of a Triangle
- Congruent Triangles

Triangles are the most common figure in geometry. In this chapter, the basic facts and relationships of triangles will be reviewed. Chapters 8 and 9 will cover several real-world applications and important theorems of triangles. Chapter 6 will review the triangle measurements of area and perimeter.

GLOSSARY

POLYGON a closed figure in a plane formed by joining segments at their endpoints

SIDE of a polygon is any one of the segments that make up the polygon

VERTEX of a polygon is any one of the common endpoints that make up the polygon. The plural of *vertex* is *vertices*.

REGULAR POLYGON a polygon in which all of the sides and all of the angles are congruent

TRIANGLE a polygon with three sides

TRIANGLE CLASSIFICATION

Triangles are classified, or grouped, in two different ways. One classification distinguishes among the sides, and another by the angles. For a triangle, you can have all three sides congruent (equal measure), or two sides congruent, or no sides congruent. Congruent sides and congruent angles of triangles are often marked as in the following figure.

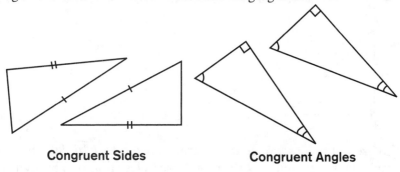

Congruent Sides   **Congruent Angles**

The following diagram shows the classification names when grouping by sides.

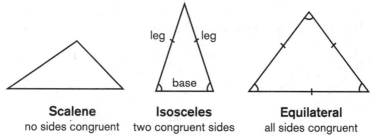

Scalene	**Isosceles**	**Equilateral**
no sides congruent	two congruent sides	all sides congruent

Note that isosceles triangles have two sides congruent, called the *legs*, and also two angles congruent, called the *base angles*. The non-congruent side is called the *base*. Equilateral triangles have all sides and all angles congruent. Each of the angles in an equilateral triangle has measure of 60°.

In Chapter 2, angle classification was reviewed. The classification of triangles according to angle measure is shown in the following figure.

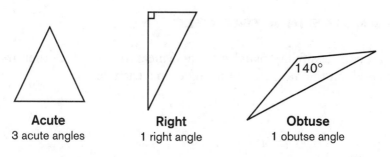

Acute	**Right**	**Obtuse**
3 acute angles	1 right angle	1 obutse angle

Be careful when classifying triangles by angle measure; notice that even though right triangles and obtuse triangles each have two acute angles, their classification is not affected by these angles. Acute triangles have all THREE acute angles.

Example:
Classify this triangle by sides and angles.

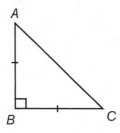

To group by sides, notice that there are two sides ($\overline{AB}$, $\overline{BC}$) that are congruent. The side classification is isosceles. To group by angles, note that there is a right angle in this triangle. So, the classification is right isosceles.

Example:
Classify this triangle by sides and angles.

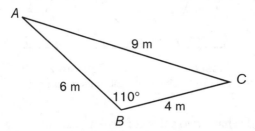

Because no sides are marked as congruent in this figure, the classification by sides is scalene. There is one angle greater than 90° ($\angle B$ is 110°); therefore, the angle classification is obtuse. This triangle is obtuse scalene.

THE ANGLES IN A TRIANGLE

In addition to classifying triangles by angle measure, there are other important facts about the measure of the angles in a triangle.

RULE BOOK
The sum of the measure of the angles in a triangle is equal to 180°.

Example:
Given the following triangle, what is the measure of $\angle XYZ$?

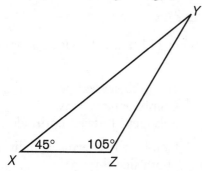

Since the measure of the other two angles are given as 105° and 45°, $\angle XYZ = 180 - (105 + 45)$, so the measure of $\angle XYZ = 180 - 150 = 30°$.

Example:
What is the measure of $\angle MNO$ in the triangle shown?

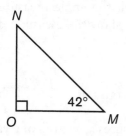

The measure of $\angle NMO$ is given, and $\angle NOM$ is a right angle, as indicated by the small box drawn in the interior. The measure of $\angle MNO = 180 - (90 + 42) = 180 - 132 = 48°$.

 ## SHORTCUT
The two acute angles in every right triangle are complementary; their sum will be 90°. In the last example, the measure of $\angle MNO$ could have been found by 90 (the measure of $\angle NMO$), which is 90 − 42 = 48°.

Example:
Use the following diagram of the triangle to find the missing angle measures.

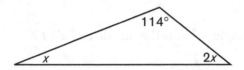

In the diagram, the three angles are 114, x and $2x$. Use algebra to solve for the variable x:

$x + 2x + 114 = 180$	Set up the equation.
$3x + 114 = 180$	Combine like terms.
$3x + 114 - 114 = 180 - 114$	Subtract 114 from both sides.
$\dfrac{3x}{3} = \dfrac{66}{3}$	Combine like terms, and divide both sides by 3.
$x = 22$	

Use the value of 22 for x, and substitute in to find the value of the missing angle measures:

The angles are $114°$, $22°$, and $2 \times 22 = 44°$.

Sometimes, problems are presented in which the angle measures are described as a ratio. In these types of problems, show the measures as the ratio factor times a variable.

Example:
In a triangle, the angle measures are in the ratio of $2 : 2 : 5$. What is the measure of the angles?
Draw the triangle, showing the angle measures as factors of a variable, such as x:

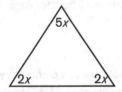

Use algebra to solve for the variable:

$2x + 2x + 5x = 180$	Set up the equation.
$\dfrac{9x}{9} = \dfrac{180}{9}$	Combine like terms, and divide both sides by 9.
$x = 20$	

Use the value of 20 to substitute in to find the measure of the angles:

$20 \times 2 = 40°, 20 \times 2 = 40°, 20 \times 5 = 100°$.

When using algebra to solve geometric problems, be clear on what the problem is asking for. Sometimes, the value of the variable is requested, and other times, such as in the last two examples, the measure of the angles is requested.

Example:
If the angle measures in a triangle are in the ratio of 1 : 2 : 3, what is the angle classification of the triangle?
The angle measures can be represented by x, $2x$, and $3x$. Use algebra to find the angle measures:

$x + 2x + 3x = 180$	Set up the equation.
$\dfrac{6x}{6} = \dfrac{180}{6}$	Combine like terms, and divide both sides by 6.
$x = 30°$	

Substitute in the value of 30 for x to get the angle measures. One of the angles measures 30°, the second angle measures $2 \times 30 = 60°$, and the third angle is $3 \times 30 = 90°$. So the triangle classification, classified by angle, is a right triangle.

GLOSSARY

EXTERIOR ANGLE an angle formed by extending one of the sides of the triangle. This extended side and one of the other original sides of the triangle define an exterior angle.

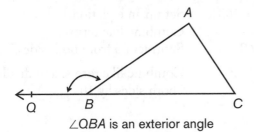

$\angle QBA$ is an exterior angle

RULE BOOK

The sum of all of the exterior angles in a triangle, or any polygon, is equal to 360°.

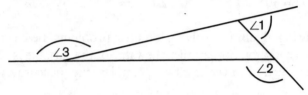

measures of ∠1 + ∠2 + ∠3 = 360°

Example:

The following diagram shows the exterior angles to a triangle. In the diagram, what is the value of *x*?

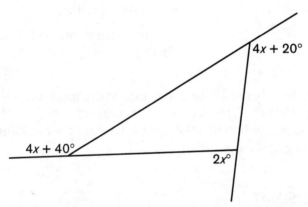

The sum of the measure of these exterior angles is equal to 360°. Use algebra:

$2x + 4x + 40 + 4x + 20 = 360$	Set up an equation.
$10x + 60 = 360$	Combine like terms.
$10x + 60 - 60 = 360 - 60$	Subtract 60 from both sides.
$\dfrac{10x}{10} = \dfrac{300}{10}$	Combine like terms, and divide both sides by 10.
$x = 30°$	

RULE BOOK

The measure of an exterior angle to a triangle is equal to the sum of the measure of the two remote interior angles of the triangle.

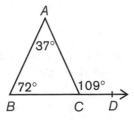

In the figure above, the measure of ∠DCA is equal to the measure of ∠CAB and ∠CBA. So, the measure of ∠DCA = 37 + 72 = 109°.

Example:
Given △XYZ, with the exterior angle shown, what is the measure of ∠XYZ?

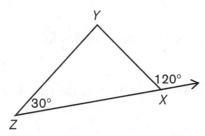

The exterior angle, whose measure is 120°, is equal to the sum of the angle of measure 30°, and ∠XYZ. The measure of ∠XYZ = 120 − 30 = 90°.

Example:
What is the value of the variable x in the following figure of a triangle with the exterior angle shown?

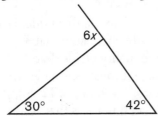

Use algebra:

$6x = 30 + 42$ Set up the equation.

$\dfrac{6x}{6} = \dfrac{72}{6}$ Combine like terms, and divide both sides by 6.

$x = 12°$

THE SIDES OF A TRIANGLE

There is an important fact about the measure of the sides of a triangle.

RULE BOOK

The sum of the measure of any two sides of a triangle is always greater than the measure of the remaining side.

Example:
Can the measures of 7 inches, 7 inches, and 15 inches form a triangle?
These side measures are not possible. Add the measures of each pair, and then ensure that this sum is greater than the third side. This must work for each pair. Adding $7 + 7 = 14$, and $14 < 15$, so these measures are not possible.

Example:
Can the measures of 7 inches, 9 inches, and 15 inches form a triangle?
Test the sum of each pair; $7 + 9 = 16$, and $16 > 15$; $7 + 15 = 22$, and $22 > 9$; $9 + 15 = 24$, and $24 > 7$. All three pairs pass the test. These measures can form a triangle.

SHORTCUT

The range of possible measures for a side of a triangle, when given the other two sides, is between the sum of the two sides, and the difference of the two sides.

Example:
If two sides of a triangle are 3 and 9, what is the range of values possible for the third side of the triangle?
The sum of any two sides must be greater than the third side. Use the variable x for the third side and then use the inequality that $3 + x > 9$. Subtracting 3 from both sides of the inequality yields that $x > 6$. The third side must also be less than 12, since it must be smaller than the sum of 3 and 9. Set up this inequality, $x < 3 + 9$, or $x < 12$. So, the range of possible measures is greater than 6 but less than 12.

RULE BOOK

In every triangle, the longest side is opposite the largest angle, and the shortest side is opposite the smallest angle.

    In an isosceles triangle, the congruent sides, called the *legs*, are opposite the congruent angles, called the *base angles*.

ALTITUDE AND MEDIAN OF A TRIANGLE

 GLOSSARY

ALTITUDE of a triangle is a segment perpendicular to one of the sides of the triangle. The length of the altitude extends from the side to the vertex opposite to the side in question. There are three altitudes in every triangle.

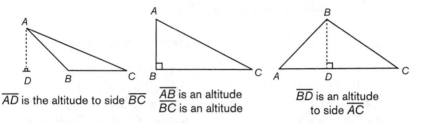

$\overline{AD}$ is the altitude to side $\overline{BC}$

$\overline{AB}$ is an altitude
$\overline{BC}$ is an altitude

$\overline{BD}$ is an altitude
to side $\overline{AC}$

MEDIAN of a triangle is a segment whose endpoints are a vertex of the triangle and the midpoint of the opposite side

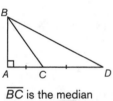 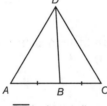 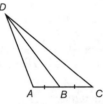

$\overline{BC}$ is the median
to side $\overline{AD}$

$\overline{DB}$ is the median
to side $\overline{AC}$

$\overline{DB}$ is the median
to side $\overline{AC}$

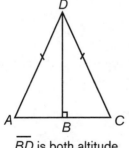 SHORTCUT

In an isosceles triangle, the median to the base is the same as the altitude to the base.

In a right triangle, two of the altitudes are sides of the triangle, the sides that form the right angle.

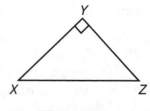

$\overline{BD}$ is both altitude
and median to side $\overline{AC}$

$\overline{XY}$ and $\overline{YZ}$ are
altitudes since
they are perpendicular

Example:
Given the following triangle, and the fact that $\overline{YW}$ is the median to $\overline{XZ}$, what is the length of $\overline{XZ}$?

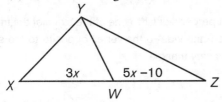

Since $\overline{YW}$ is a median to $\overline{XZ}$, it bisects $\overline{XZ}$. Use algebra:

$3x = 5x - 10$	Set up the equation.
$3x + 10 = 5x - 10 + 10$	Add 10 to both sides.
$3x + 10 = 5x$	Combine like terms.
$3x - 3x + 10 = 5x - 3x$	Subtract $3x$ from both sides.
$\frac{10}{2} = \frac{2x}{2}$	Combine like terms, and divide both sides by 2.
$5 = x$	

Use the value of 5 to find the measure of $\overline{XZ}$: $(3 \times 5) + (5 \times 5 - 10)$ = $15 + 25 - 10 = 40 - 10 = 30$.

Example:
In the following triangle, which of the following statements is true?

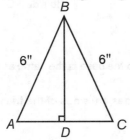

measure of $\overline{AC} = 4"$

a. $\triangle ABC$ is an isosceles triangle.
b. The median to $\overline{AC}$ is the same segment as the altitude to $\overline{AC}$.
c. $\triangle ABD$ is a right triangle.
d. $\triangle ABC$ is an acute triangle.
e. All of the above are true.

The correct response is choice **e**. All of the three choices are true. Choice **a** is true because two of the sides have equal measure. Choice **b** is true because in an isosceles triangle, the median of the base is the same as the altitude to the base. Choice **c** is true because this triangle contains a 90° angle. Choice **d** is true because for this triangle, all of the angles are less than 90°.

CONGRUENT TRIANGLES

Recall from Chapter 2 that two figures are congruent if they have the same shape and the same size. Look at the following congruent triangles. They have the same size and the same shape. Imagine lifting one of them and placing it on top of the other. Both sides and vertices could be matched up.

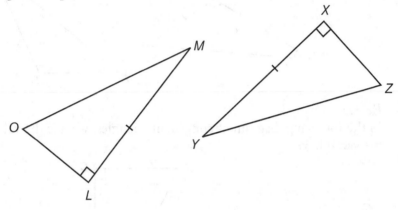

These matching sides and vertices are referred to as corresponding sides and corresponding vertices. In the diagram, *L* would correspond to *X*; *M* would correspond to *Y*; and *O* would correspond to *Z*.

RULE BOOK

Corresponding sides of congruent triangles are congruent.

Corresponding angles of congruent triangles are congruent.

When naming congruent triangles, the corresponding vertices are listed in the same order. This rule is followed so that when given the statement △*ABC* ≅ △*DEF*, you know the corresponding parts even without a diagram. In the above congruence, $\overline{AB}$ is congruent to $\overline{DE}$, and ∠*BCA* ≅ ∠*EFD*. This is just two of six corresponding congruent parts that are defined by the triangle congruence statement. The following triangles are a possible picture for the congruence statement given.

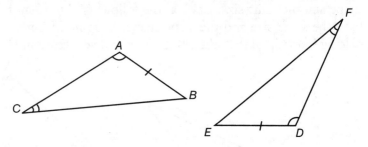

Example:

In the following diagram of congruent triangles, what is the value of the variable *x*?

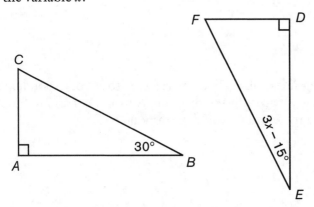

In the diagram ∠*ABC* ≅ ∠*DEF*. The measure of ∠*ABC* = 30°. Use algebra to solve for the variable *x*:

$3x - 15 = 30$	Set up the equation.
$3x - 15 + 15 = 30 + 15$	Add 15 to both sides.
$\dfrac{3x}{3} = \dfrac{45}{3}$	Combine like terms, and divide both sides by 3.
$x = 15°$	

Example:
Given that $\triangle ABC \cong \triangle LMN$, which of the following is true?
a. $\angle BAC \cong \angle MNL$
b. $\overline{LN} \cong \overline{AB}$
c. $\angle CBA \cong \angle NML$
d. $\overline{MN} \cong \overline{AB}$
e. $\angle ABC \cong \angle LNM$

Choice **c** is the only true statement. Use the congruence statement to establish the congruent vertices. Vertex A corresponds to L, B corresponds to M, and C corresponds to N.

Example:
Given the following congruent triangles, what is the value of the variable x?

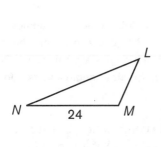 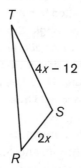

$\triangle RST \cong \triangle LMN$; $\overline{MN} \cong \overline{ST}$. Use algebra to solve for the variable x:

$4x - 12 = 24$	Set up the equation.
$4x - 12 + 12 = 24 + 12$	Add 12 to both sides.
$\frac{4x}{4} = \frac{36}{4}$	Combine like terms, and divide both sides by 4.
$x = 9$	

Triangle congruence can be assured without knowing all the measures of all of the sides and angles. There are special cases for testing congruence that only involve three corresponding parts of the triangles. These are mathematical methods that have been proven and are always true. The proof of these methods is beyond the scope of this book. However, knowledge of the methods is important in order to test triangle congruence.

RULE BOOK

There are four established methods of testing triangle congruence.

SSS (Side-Side-Side) method states that if the three sides of one triangle are congruent to the three sides of another triangle, the triangles are congruent.

SAS (Side-Angle-Side) method states that if two sides and the angle that they form (the included angle) are congruent to two sides and the included angle of another triangle, the triangles are congruent. The angle MUST be between the sides for this method to hold true.

ASA (Angle-Side-Angle) method states that if two angles and the side between them (the included side) are congruent to two angles and the included side of another triangle, then the triangles are congruent. The side MUST be between the angles for this method to hold true.

AAS (Angle-Angle-Side) method states that if two angles and a side not included are congruent to two angles and the corresponding side not included, the triangles are congruent. One of the angles MUST be between the other angle and side for this method to hold true.

When you are asked to determine triangle congruence, be careful. Note that AAA (Angle-Angle-Angle) and SSA (Side-Side-Angle) are NOT methods that can be used to establish congruence! In addition, if a test question asks whether two triangles are congruent, you cannot assume congruence just because they appear congruent. You may encounter test questions that give a drawing of two triangles and limited information about the triangles. The methods above may be needed to establish whether the two triangles are congruent.

Example:
In the following two triangles, what method could be used to establish congruence, based on the information given?

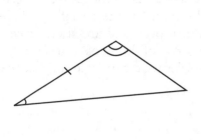

The correct response is the ASA method. In the diagram, the congruent parts that are marked are two angles, and the side that is in between these angles. If information is given in words, label the triangles with the congruence symbols and then use the figure to determine the correct method. In triangle congruence problems, look for shared sides. If two triangles share a common side, then this is a side that can be used to establish congruence.

Example:
Given that $\overline{AD}$ is the perpendicular bisector of △ABC, shown below, what method can be used to establish congruence of △ADC and △ADB?

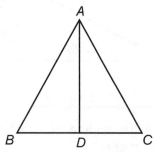

Draw in the perpendicular bisector, including the marking for the right angle and the congruent segments. The two triangles share the common side of $\overline{AD}$. This is marked with an "X."

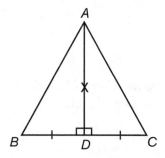

From the drawing, it is apparent that the method is SAS, since it is two congruent sides and the included angle between them.

Triangles may also overlap. It is helpful in these cases to show the triangles separated, as in the following diagram. When separated, mark all congruent parts.

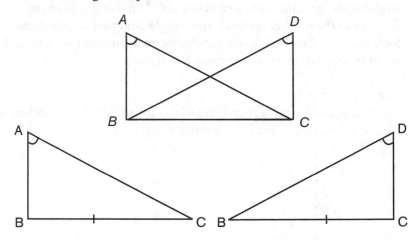

Example:
Given the following ΔGHI and ΔJIH, what method could be used to establish that the two triangles are congruent?

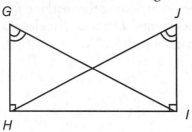

Separate the triangles, and transfer the markings. Mark the shared segment, that is $\overline{HI}$, as congruent in the two triangles.

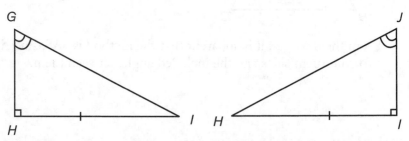

From the drawing, the method is AAS. The marked congruence is an angle, then another angle, which is the right angle, and then a side, which is the shared side. One of the angles is in between the other angle and side.

EXTRA HELP

The website *www.math.com* has an interactive lesson on triangles. On the left sidebar, click on *Geometry*. Then, under *Polygons*, click on *Triangles*. There is a series of interactive lessons, followed by a short interactive quiz. Another website is http://library.thinkquest.org/20991/geo/index.html. Upon reaching this site, click on *Congruent Triangles*, located on the left-hand list. For more information and practice on working with triangles, reference the book *Geometry Success in 20 Minutes a Day*, published by LearningExpress.

TIPS AND STRATEGIES

When working with triangles, remember:

- Triangles are classified one way according to the angle measure and are classified a second way according to side measure.
- For a triangle to be classified as acute, all three angles must be acute angles.
- The sum of the degree measure of the angles in a triangle is 180°.
- The two acute angles in a right triangle are complementary.
- The sum of the degree measure of the three exterior angles of a triangle is 360°.
- The measure of an exterior angle to a triangle is equal to the sum of the measure of the two remote interior angles.
- An altitude to a side of a triangle is the perpendicular segment to that side.
- A median is a segment whose endpoints are a vertex of the triangle and the midpoint of the opposite side.
- Corresponding sides of congruent triangles are congruent.
- Corresponding angles of congruent triangles are congruent.
- There are four methods to establish triangle congruence: SSS, SAS, AAS, and ASA.
- The methods of AAA and SSA are not acceptable methods to establish congruence.

When you are taking a multiple-choice test, remember these tips to improve your score:

- For triangle problems that ask for a method to establish congruence, immediately eliminate any methods that are AAA, or SSA.

- For triangle problems that ask for a method to establish congruence, only use the information given. Do not make assumptions based on the appearance of the figure.

PRACTICE QUIZ

Try these 25 problems. When you are finished, review the answers and explanations to see if you have mastered this concept.

1. Which of the following side measures could NOT form a triangle?
   a. 3, 4, and 5
   b. 6, 6, and 12
   c. 12, 13, and 15
   d. 5, 5, and 5
   e. 4, 7, and 10

2. If two sides of a triangle are 6 and 10, what is the exact range of possible measure for the third side?
   a. any measure less than 6
   b. any measure greater than 10
   c. a measure between 6 and 10 only
   d. a measure between 4 and 16 only
   e. A triangle is not possible with side lengths of 6 and 10.

3. What is the classification of the following triangle?

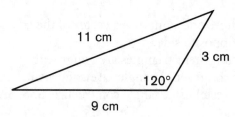

   a. acute obtuse
   b. acute scalene
   c. acute isosceles
   d. acute right
   e. obtuse scalene

4. What is the classification of the following triangle?

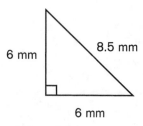

    a. acute right
    b. acute isosceles
    c. right isosceles
    d. acute scalene
    e. isosceles scalene

5. By which method are $\triangle ABC$ and $\triangle DBC$ congruent, given the figure below and the fact that $\overline{BC}$ is a perpendicular bisector of $\overline{AD}$?

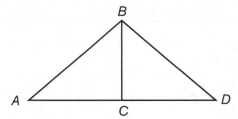

    a. SAS
    b. SSS
    c. AAS
    d. SAA
    e. SSA

6. By which method are the two triangles congruent, given the following figures?

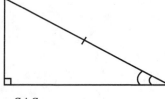 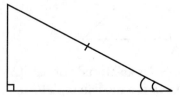

    a. SAS
    b. AAS
    c. ASA
    d. SSS
    e. SSA

7. Given the following triangle, what is the measure of ∠*SRT*?

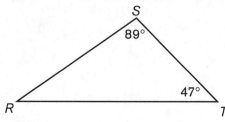

   a. 44°
   b. 224°
   c. 47°
   d. 89°
   e. 4.4°

8. What is the value of *x* in the following triangle?

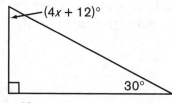

   a. 48
   b. 12
   c. 60
   d. 15
   e. 19.5

9. Which of the statements is ALWAYS true about the following triangle?

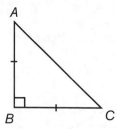

   a. $\overline{AB}$ is an altitude of △*ABC*.
   b. $\overline{AB}$ is a median of △*ABC*.
   c. △*ABC* is acute isosceles.
   d. $\overline{AC} \cong \overline{BC}$
   e. ∠*ABC* ≅ ∠*ACB*

10. Given the following diagram, which statement is true?

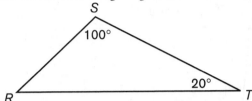

a. The measure of $\overline{ST}$ is less than the measure of $\overline{RS}$.
b. The measure of $\overline{ST}$ is greater than the measure of $\overline{RT}$.
c. The measure of $\overline{RT}$ is greater than the measure of $\overline{ST}$.
d. The measure of $\angle SRT$ is equal to 80°.
e. The measure of $\angle SRT$ is equal to 100°.

11. Given $\triangle BDC$ with exterior angles shown, what is the degree measure of $\angle ABD$?

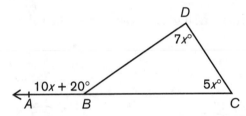

a. 120°
b. 30°
c. 93°
d. 10°
e. 12°

12. Given the two congruent triangles LMN and RST shown, what is the length of $\overline{LN}$?

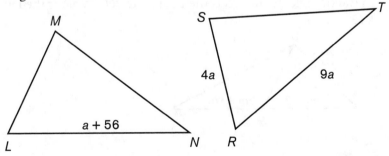

a. 7
b. 56
c. 63
d. 28
e. 91

13. Use the following diagram of a triangle with the three exterior angles shown to find the value of the variable x.

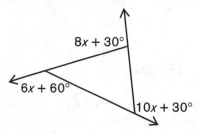

$8x + 30°$

$6x + 60°$

$10x + 30°$

   a. 90°
   b. 360°
   c. 120°
   d. 10°
   e. 18°

14. Use the figure of congruent triangles below to find the value of the variable x.

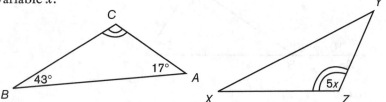

   a. 43°
   b. 10°
   c. 24°
   d. 8.6°
   e. 120°

15. In the triangle below, $\overline{US}$ is the median to $\overline{RT}$. What is the length of $\overline{RT}$?

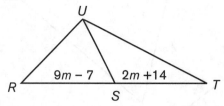

   a. 3
   b. 20
   c. 15.72
   d. 31
   e. 40

16. The ratio of the degree measure of the angles in a triangle is 2 : 3 : 5. What is the angle classification of the triangle?
   a. obtuse triangle
   b. right triangle
   c. straight triangle
   d. acute triangle
   e. isosceles triangle

17. The ratio of the degree measure of the exterior angles to a triangle is 5 : 6 : 9. What is the measure of the largest exterior angle?
   a. 162°
   b. 180°
   c. 90°
   d. 108°
   e. 262°

18. Given that $\triangle XYZ$ is an isosceles triangle where $\overline{XZ}$ is the base and $\overline{YM}$ is the median to the base, by which method below can it be established that $\triangle XYM \cong \triangle ZYM$?
   a. AAA
   b. SSA
   c. ASA
   d. SSS
   e. AAS

19. The two acute angles in a right triangle are in the ratio of 2 : 4. What is the measure of the smallest angle?
   a. 15°
   b. 30°
   c. 90°
   d. 60°
   e. 120°

20. Given the following diagram of two triangles, which methods can be used to determine congruence?

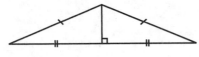

   a. SAS
   b. SSA
   c. SSS
   d. either choice **a** or **b**
   e. either choice **a** or **c**

21. What is the degree measure of an exterior angle to an equilateral triangle?

   a. 60°

   b. 180°

   c. 300°

   d. 120°

   e. 360°

22. Given the following diagram of △ABC, which of the following statements is true?

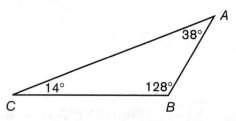

   a. △ABC is an acute scalene triangle.

   b. The measure of $\overline{BC}$ is greater than the measure of $\overline{AC}$.

   c. The measure of $\overline{AB}$ is greater than the measure of $\overline{BC}$.

   d. The measure of $\overline{AC}$ is greater than the measure of $\overline{BC}$.

   e. None of the above is true.

23. Classify the following triangle by angles and sides.

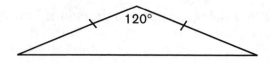

   a. obtuse isosceles

   b. right scalene

   c. right isosceles

   d. acute obtuse

   e. isosceles scalene

24. Given the following overlapping triangles, by what method can it be established that $\triangle ABC \cong \triangle DCB$?

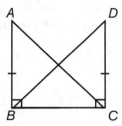

a. SSS
b. SAS
c. AAS
d. AAA
e. SSA

25. In the following figure of overlapping triangles, by what method can it be established that the two triangles are congruent?

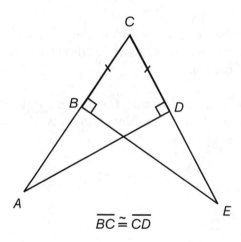

$$\overline{BC} \cong \overline{CD}$$

a. SAS
b. AAA
c. SSS
d. AAS
e. ASA

ANSWERS

Here are the answers, with detailed explanations, for the triangle practice quiz. Study the explanations for any questions that you answered incorrectly. Then, go back and try the problems again.

1. b. The measures of 6, 6, and 12 cannot form a triangle. The sum of two of the sides, that is 6 + 6 = 12, is equal to the third side. The sum of any two sides must be GREATER than the third side.

2. d. To find the exact range of possible values, remember the shortcut that the third side must be greater than the difference, that is 10 – 6 = 4. The third side must also be smaller than the sum, that is 10 + 6 = 16.

3. e. The classification is scalene; all of the sides have a different measure. The angle classification is obtuse because one of the angles, namely 120°, is obtuse.

4. c. The classification is isosceles because two of the sides have equal measure. The angle classification is "right," as shown by the small box marked in one of the angles.

5. a. Since $\overline{BC}$ is a perpendicular bisector of $\overline{AD}$, it forms right angles and bisects $\overline{AD}$, creating the two congruent segments $\overline{AC}$ and $\overline{CD}$. Draw in these congruent parts as shown:

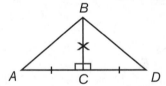

From the figure, the two triangles share the common side of $\overline{BC}$; this is the third congruent part. The correct method is SAS, because the right angle is between the congruent sides.

6. b. The method for congruence is angle-angle-side, or AAS, since one of the angles is between the other angle and side.

7. a. The sum of the degree measure of the angles in a triangle is 180°. The measure of $\angle SRT = 180 – 89 – 47 = 180 – 136 = 44°$.

8. b. The two acute angles in a right triangle are complementary. Use algebra to solve for x.

$$4x + 12 + 30 = 90 \qquad \text{Set up an equation.}$$
$$4x + 42 = 90 \qquad \text{Combine like terms.}$$
$$4x + 42 - 42 = 90 - 42 \qquad \text{Subtract 42 from both sides.}$$
$$4x = 48 \qquad \text{Combine like terms.}$$
$$\frac{4x}{4} = \frac{48}{4} \qquad \text{Divide both sides by 4.}$$
$$x = 12$$

9. a. The only true statement is that $\overline{AB}$ is an altitude of $\triangle ABC$. $\overline{AB}$ is not a median of $\triangle ABC$. The median bisects a side of the triangle. The classification of $\triangle ABC$ is right isosceles, not acute isosceles. The congruent sides are $\overline{AB}$ and $\overline{BC}$. $\overline{AC}$ is not congruent to the other sides. $\angle ABC$ and $\angle ACB$ cannot be congruent; there is at most one right angle in a triangle.

10. c. The measure of $\overline{RT}$ is greater than the measure of $\overline{ST}$ because $\angle RST$ is the largest angle, and the largest side is opposite to the largest angle in a triangle.

11. a. The degree measure of an exterior angle to a triangle is equal to the sum of the other two interior angles. Use algebra to solve for the variable x, and then find the measure of the angle.

$$10x + 20 = 5x + 7x \qquad \text{Set up an equation.}$$
$$10x + 20 = 12x \qquad \text{Combine like terms.}$$
$$10x - 10x + 20 = 12x - 10x \qquad \text{Subtract } 10x \text{ from both sides.}$$
$$20 = 2x \qquad \text{Combine like terms.}$$
$$\frac{20}{2} = \frac{2x}{2} \qquad \text{Divide both sides by 2.}$$
$$10 = x$$

Use the value of 10 for x to find the measure of $10x + 20$:
$(10 \times 10) + 20 = 100 + 20 = 120°$.

12. c. The congruent sides are $\overline{LN}$ and $\overline{RT}$, based on the given congruence statement. Use algebra to solve for the variable a, and then find the length of $\overline{LN}$.

$$a + 56 = 9a \qquad \text{Set up an equation.}$$
$$a - a + 56 = 9a - a \qquad \text{Subtract } a \text{ from both sides.}$$
$$56 = 8a \qquad \text{Combine like terms.}$$
$$\frac{56}{8} = \frac{8a}{8} \qquad \text{Divide both sides by 8.}$$
$$7 = a$$

Substitute in the value of 7 to find the length of $\overline{LN} = a + 56$; $7 + 56 = 63$.

13. d. The sum of the degree measure of the exterior angles to a triangle is 360°. Use algebra to solve for x.

$(8x + 30) + (6x + 60) + (10x + 30) = 360$	Set up an equation.
$24x + 120 = 360$	Combine like terms.
$24x + 120 - 120 = 360 - 120$	Subtract 120 from both sides.
$24x = 240$	Combine like terms.
$\frac{24x}{24} = \frac{240}{24}$	Divide both sides by 24.
$x = 10$	

14. c. $\angle XZY$ is congruent to $\angle ACB$ by using the marking that shows these angles to be congruent. Their measures are equal. The measure of $\angle ACB$ is 120°, because the sum of the angle measures in a triangle is 180°, and $180 - 43 - 17 = 120$. Use algebra to solve for x.

$5x = 120$	Set up an equation.
$\frac{5x}{5} = \frac{120}{5}$	Divide both sides by 5.
$x = 24$	

15. e. The median $\overline{US}$ bisects $\overline{RT}$. From this information, $\overline{RS} \cong \overline{ST}$ and their measures are equal. Use algebra to solve for m, and then find the length of $\overline{RT}$.

$9m - 7 = 2m + 14$	Set up an equation.
$9m - 2m - 7 = 2m - 2m + 14$	Subtract $2m$ from both sides.
$7m - 7 = 14$	Combine like terms.
$7m - 7 + 7 = 14 + 7$	Add 7 to both sides.
$\frac{7m}{7} = \frac{21}{7}$	Combine like terms, and divide both sides by 7.
$m = 3$	

Use the value of 3 to find the length of $\overline{RT}$: $9m - 7 + 2m + 14$; $(9 \times 3) - 7 + (2 \times 3) + 14 = 27 - 7 + 6 + 14 = 40$.

16. b. The sum of the degree measures of the angles in a triangle is 180°. Because ratios are given for angle measure, express these measures as $2x$, $3x$, and $5x$ respectively. Use algebra to solve for x, and then find the measure of the angles to find the classification.

$2x + 3x + 5x = 180$ Set up an equation.

$\frac{10x}{10} = \frac{180}{10}$ Combine like terms, and divide both sides by 10.

$x = 18$

Using the value of 18 for x, the angle measures are $2 \times 18 = 36$, $3 \times 18 = 54$, and $5 \times 18 = 90$. The angle classification is a right triangle because one of the angles is 90°.

17. a. The sum of the degree measures of the exterior angles in a triangle is 360°. Because ratios are given for angle measure, express these measures as $5x$, $6x$, and $9x$ respectively. Use algebra to solve for x, and then find the measure of the largest angle.

$5x + 6x + 9x = 360$ Set up an equation.

$\frac{20x}{20} = \frac{360}{20}$ Combine like terms, and divide both sides by 20.

$x = 18$

The largest angle is 9 times x, which is 162°.

18. d. Draw the isosceles triangle with given information:

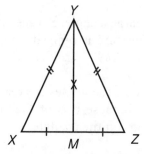

The fact that $\triangle XYZ$ is an isosceles triangle means that $\overline{XY}$ and $\overline{YZ}$ are congruent. Because $\overline{YM}$ is the median to the base, $\overline{XM} \cong \overline{MZ}$. The two triangles share a common side $\overline{YM}$, which is the third necessary congruent part. The method is side-side-side, or SSS.

19. b. The degree measure of the two acute angles in a right triangle add up to 90°. A ratio is given, so use the expressions $2x$ and $4x$ to represent the acute angles. Use algebra to solve for x, and then find the measure of the smallest angle, which is 2 times x.

$2x + 4x = 90$	Set up an equation.
$\dfrac{6x}{6} = \dfrac{90}{6}$	Combine like terms, and divide both sides by 6.
$x = 15$	

The smallest angle is 2 times 15, which is 30°.

20. e. There are two possible methods to establish congruence, that is SAS or SSS. The two triangles share a common side and the other two sides are marked as congruent. In addition, the small box in the interior of the angle indicates these segments are perpendicular, so both triangles have the congruent angle of 90°. Note that even though the diagram shows congruent parts of side-side-angle, or SSA, this is NOT an acceptable method to use to establish congruence.

21. d. The angles in an equilateral triangle are all congruent. Since their measure must add up to 180°, each interior angle is 60°. The measure of an exterior angle is equal to the sum of the two remote interior angles, so the measure of an exterior angle in this case is $60 + 60 = 120°$.

22. d. The longest side is opposite to the largest angle. Because $\angle ABC = 128°$, the side with the greatest measure is opposite this angle, $\overline{AC}$. Note also that $\overline{BC} > \overline{AB}$ because $\angle BAC > \angle BCA$.

23. a. The triangle is obtuse because one of the angles is greater than 90°. The classification is isosceles because two of the sides are congruent, as shown by the markings on the triangle.

24. b. First, separate the two triangles, and mark all congruent parts. Remember that $\overline{BC}$ is a shared side, so should be marked as congruent on both:

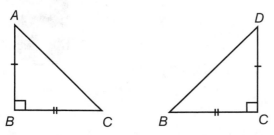

From the new diagram, the method is clearly side-angle-side, or SAS. The angle is between the two congruent sides.

25. e. First, separate the two triangles, and mark all congruent parts. Remember that $\angle BCD$ is a shared angle, so it should be marked as congruent on both:

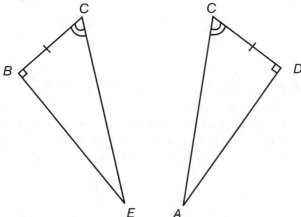

From the new diagram, the method is angle-side-angle, or ASA. The side is between the two congruent angles.

5

Quadrilaterals and Circles

Chapter 4 introduced the basic facts about triangles. This chapter concentrates on the other two most common closed figures—the quadrilateral and the circle. Start your study of these figures by taking the ten-question benchmark quiz. Grade yourself and read through the answer explanations to determine how much knowledge you already have on these topics.

BENCHMARK QUIZ

1. Which of the following statements is FALSE?
   a. All squares are rhombuses.
   b. All trapezoids are quadrilaterals.
   c. All trapezoids have two congruent sides.
   d. All squares are rectangles.
   e. All rhombuses are parallelograms.

2. Which of the following statements is ALWAYS true?

    a. The diagonals of a rectangle are congruent.
    b. The diagonals of a rhombus are congruent.
    c. The diagonals of a parallelogram are congruent.
    d. The diagonals of a trapezoid are congruent.
    e. All of the above are always true.

3. Using rectangle $QRST$ below, what is the measure of $\overline{RS}$?

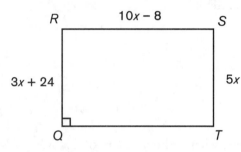

$5x + 24 = 5x$

$x = 12$

$(20$

112

    a. 112
    b. 120
    c. 12
    d. 8
    e. 72

4. In the following parallelogram, what is the value of the variable x?

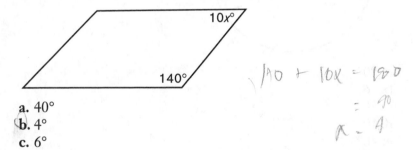

$140 + 10x = 180$

$= 40$

$x = 4$

    a. 40°
    b. 4°
    c. 6°
    d. 22°
    e. 60°

5. In the following rectangle *ABCD*, the measure of $\overline{BD}$ = 84, and the measure of $\overline{AE}$ = 6*x*. What is the value of the variable *x*?

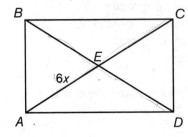

12x = 84

6x = 42

x = 7

a. 14
b. 7
c. 84
d. 28
e. 3.5

6. Given the following circle, what is the measure of ∠*ABC*?

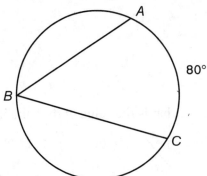

80°

a. 80°
b. 160°
c. 50°
d. 100°
e. 40°

7. In the following circle, the measure of $\angle AOB = 70°$. What is the measure of $\overset{\frown}{CB}$?

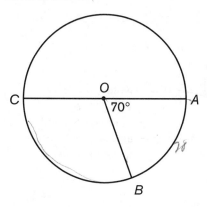

a. 70°
b. 140°
c. 145°
d. 110°
e. 115°

8. In the following circle, what is the measure of $\overline{DC}$?

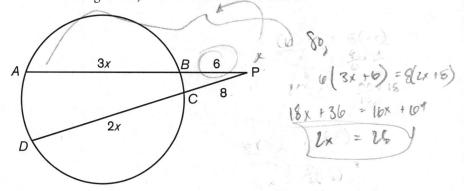

a. 36
b. 8
c. 28
d. 6
e. 42

9. In the following circle, what is the measure of $\overline{AC}$?

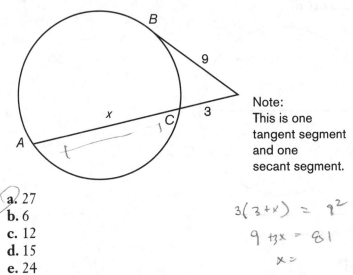

Note:
This is one
tangent segment
and one
secant segment.

a. 27
b. 6
c. 12
d. 15
e. 24

$3(3+x) = 9^2$

$9 + 3x = 81$

$x =$

10. In the following circle, what is the measure of $\overline{DB}$?

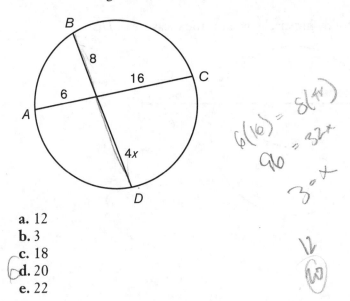

a. 12
b. 3
c. 18
d. 20
e. 22

$6(16) = 8(4x)$

$96 = 32x$

$3 = x$

BENCHMARK QUIZ SOLUTIONS

Carefully check your answers, and read through the answer explanations. Grade yourself, and then follow the suggestions given under *Benchmark Quiz Results*.

1. c. Not all trapezoids have two congruent sides. Only isosceles trapezoids have two congruent sides. Every square is both a rhombus and a rectangle. A trapezoid is a quadrilateral; it has four sides. Every rhombus is a parallelogram.

2. a. The diagonals of a rectangle are always congruent. The only time a rhombus has congruent diagonals is when the rhombus is a square. The only time a parallelogram has congruent diagonals is when it is a rectangle. The only time a trapezoid has congruent diagonals is when it is an isosceles trapezoid.

3. a. The opposite sides of a rectangle are congruent. Use this fact and algebra to find the value of the variable x. Use this value to calculate the length of $\overline{RS}$.

$5x = 3x + 24$	Set up an equation.
$5x - 3x = 3x + 24 - 3x$	Subtract $3x$ from both sides.
$2x = 24$	Combine like terms.
$\frac{2x}{2} = \frac{24}{2}$	Divide both sides by 2.
$x = 12$	

Use this value of 12 to find the length of $\overline{RS}$:
$10x - 8 = (10 \times 12) - 8 = 120 - 8 = 112$.

4. b. The measure of the consecutive angles in a parallelogram add up to 180°. Use algebra:

$10x + 140 = 180$	Set up an equation.
$10x + 140 - 140 = 180 - 140$	Subtract 140 from both sides.
$10x = 40$	Combine like terms.
$\frac{10x}{10} = \frac{40}{10}$	Divide both sides by 10.
$x = 4$	

5. b. The diagonals of a rectangle are congruent; they also bisect each other. The measure of $\overline{AC}$ = $6x + 6x = 12x$. Use algebra:

$12x = 84$ Set up an equation.

$\dfrac{12x}{12} = \dfrac{84}{12}$ Divide both sides by 12.

$x = 7$

6. e. The measure of an inscribed angle is one-half the measure of the arc that it intercepts. The measure of $\angle ABC = 80 \div 2 = 40°$.

7. d. $\overline{AC}$ is a diameter of the circle, therefore $\overset{\frown}{CA} = 180°$. The measure of $\overset{\frown}{AB}$ is equal to the measure of the central angle shown, that is 70°. The measure of $\overset{\frown}{CB} = 180 - 70 = 110°$.

8. c. For the two secants shown, the lengths are in the relationship of (outer piece) times (whole segment) = (outer piece) times (whole segment). Use algebra to find the value of x, and then find the length of $\overline{DC}$.

$6(3x + 6) = 8(2x + 8)$ Set up an equation.

$(6 \times 3x) + (6 \times 6) = (8 \times 2x) + (8 \times 8)$ Use the distributive property.

$18x + 36 = 16x + 64$ Multiply in the parentheses.

$18x + 36 - 16x = 16x + 64 - 16x$ Subtract $16x$ from both sides.

$2x + 36 = 64$ Combine like terms.

$2x + 36 - 36 = 64 - 36$ Subtract 36 from both sides.

$2x = 28$ Combine like terms.

$\dfrac{2x}{2} = \dfrac{28}{2}$ Divide both sides by 2.

$x = 14$

The length of $\overline{DC}$ is $2x$, so the length is $2 \times 14 = 28$.

9. e. For the secant and tangent shown, the lengths are in the relation-ship of (outer piece) times (whole segment) = (tangent)^2. Use alge-bra to find the value of x, which is the length of $\overline{AC}$.

$3(x + 3) = 92$	Set up an equation.
$(3 \times x) + (3 \times 3) = 9^2$	Use the distributive property.
$3x + 9 = 81$	Multiply.
$3x + 9 - 9 = 81 - 9$	Subtract 9 from both sides.
$3x = 72$	Combine like terms.
$\frac{3x}{3} = \frac{72}{3}$	Divide both sides by 3.
$x = 24$	This is the length of $\overline{AC}$.

10. d. Multiply the two segments of one chord together. This value is equal to the product of the two segments of the other chord shown. Use algebra to find the value of the variable x. Use this value to find the length of $\overline{DB}$.

$8(4x) = 6(16)$	Set up an equation.
$32x = 96$	Multiply.
$\frac{32x}{32} = \frac{96}{32}$	Divide both sides by 32.
$x = 3$	

Use this value of 3 to find the length of $\overline{DB}$: $4x + 8 = (4 \times 3) + 8 = 12 + 8 = 20$.

BENCHMARK QUIZ RESULTS

If you answered 8–10 questions correctly, quadrilaterals and circles are strengths for you. Read over the chapter, concentrating on the areas where you were weak. Refer to the sidebars to pick up techniques and vocabulary. The benchmark quiz did not cover all the material in the chapter; work through the end of the Practice Quiz to ensure that you have a full under-standing of these topics.

If you answered 4–7 questions correctly, spend some time on this chap-ter. Read through the entire chapter. Pay attention to the sidebars that refer you to more in-depth practice, hints, and shortcuts. Work through the Practice Quiz at the end of the chapter to check your progress. The answer explanations will help you through any areas that still need clarification. Visit the suggested websites for additional practice.

If you answered 1–3 questions correctly, you need a thorough review of quadrilaterals and circles. First, carefully read this chapter and concentrate on the sidebars and all figures. Be sure you know all the rules outlined in the Rule Book sidebars, and all of the necessary vocabulary. Go to the suggested websites in the Extra Help sidebar in this chapter and do extended practice. For further study, refer to the book, *Geometry Success in 20 Minutes a Day*, Lesson 10, published by LearningExpress.

JUST IN TIME LESSON— QUADRILATERALS AND CIRCLES

Topics in this chapter include:

- Quadrilateral Classification
- The Angles in a Quadrilateral
- The Sides and Diagonals of Quadrilaterals
- The Parts of a Circle
- Angles and Arcs Related to Circles and Their Measure
- Segment Measures as Related to Circles

This chapter examines the basic facts and relationships of quadrilaterals and circles. Chapter 6 will review the measurements of area and perimeter. Chapter 7 will cover surface area and volume of three-dimensional figures.

QUADRILATERAL CLASSIFICATION

GLOSSARY
QUADRILATERAL a four sided polygon

Four-sided polygons are called *quadrilaterals* and, like triangles, there are classifications for quadrilaterals. A quadrilateral with one pair of parallel sides (bases) is called a *trapezoid*.

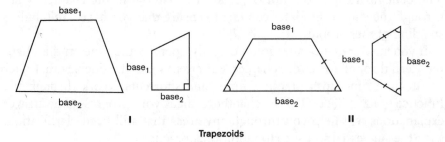

Trapezoids

In an *isosceles trapezoid*, the non-base sides are congruent. An example can be found in figure II in the above graphic. Because the parallel bases are not the same length in a trapezoid, we call these bases b_1 and b_2.

A quadrilateral with two pairs of parallel sides is called a *parallelogram*. The two sets of opposite sides that are parallel are congruent in a parallelogram, as shown in the following figures:

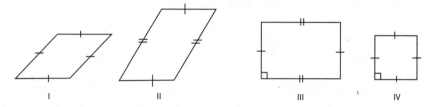

Parallelograms are broken down into further sub-groups.

A *rectangle* is a parallelogram with four right angles. Refer to figure III, above.

A *rhombus* is a parallelogram with four congruent sides. Refer to figure I, above.

A *square* is a parallelogram with both four right angles and four congruent sides. A square is a rhombus, a rectangle, a parallelogram, and a quadrilateral. Refer to figure IV, above.

This Venn diagram may help you to understand the classification of four-sided figures:

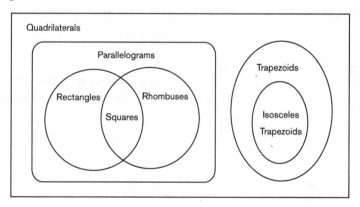

For example, in the Venn diagram a rectangle is a parallelogram and a quadrilateral since it is contained in the other classifications. A square is a rectangle, rhombus, parallelogram, and a quadrilateral. A trapezoid is not a parallelogram; it is not contained in the parallelogram sub-group.

Example:
Which of the following statements is FALSE?
a. A square is a parallelogram.
b. A rhombus is a quadrilateral.
c. A trapezoid is a quadrilateral.
d. A rectangle is a trapezoid.
e. A square is a rhombus.

The choice that is false is choice **d**. A rectangle is NOT a trapezoid. Refer to the Venn diagram. The rectangle is not contained in the trapezoid section. All of the other choices are true.

THE ANGLES IN A QUADRILATERAL

 RULE BOOK

The sum of the measure of the angles in a quadrilateral is equal to 360°.
The sum of the measures of any two consecutive angles in a parallelogram is equal to 180°.

There are special rules that govern the angles in some quadrilaterals:

 RULE BOOK

In a parallelogram, opposite angles are congruent. This is true for all parallelograms, including rectangles, rhombuses, and squares.

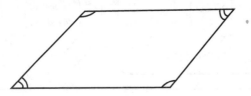

In an isosceles trapezoid, the base angles are congruent.

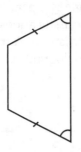
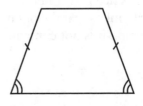

Example:
In the parallelogram below, what is the measure of ∠*ABC*?

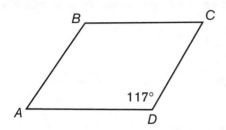

The angle is congruent to the angle opposite, which has measure of 117°. The measure of ∠*ABC* = 117°.

Example:
In the isosceles trapezoid shown, what is the measure of ∠*WXY*?

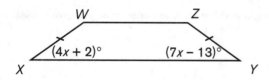

The base angles are congruent. Use algebra to solve for the variable *x*, and then find the measure of ∠*WXY*:

$4x + 2 = 7x - 13$	Set up an equation.
$4x + 2 - 4x = 7x - 13 - 4x$	Subtract $4x$ from both sides.
$2 = 3x - 13$	Combine like terms.
$2 + 13 = 3x - 13 + 13$	Add 13 to both sides.
$15 = 3x$	Combine like terms.
$\dfrac{15}{3} = \dfrac{3x}{3}$	Divide both sides by 3.
$5 = x$	

Use the value of 5 to find the measure of ∠*WXY*:
$4x + 2 = (4 \times 5) + 2 = 20 + 2 = 22°$.

GLOSSARY

EXTERIOR ANGLE an angle formed by extending one of the sides of the quadrilateral. This extended side and the adjacent side of the original quadrilateral define an exterior angle.

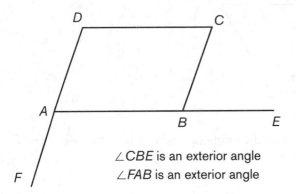

∠*CBE* is an exterior angle

∠*FAB* is an exterior angle

 RULE BOOK

An exterior angle and the adjacent interior angle form a linear pair.

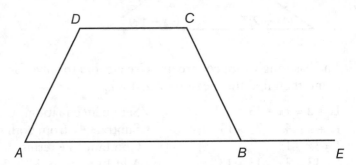

measure ∠*CBA* + measure ∠*CBE* = 180°

(m ∠*CBA*) + (m ∠*CBE*) = 180°

The sum of all of the exterior angles in a quadrilateral, or any polygon, is equal to 360°.

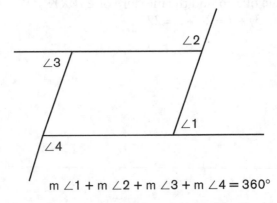

m ∠1 + m ∠2 + m ∠3 + m ∠4 = 360°

SHORTCUT

Because opposite angles in a parallelogram are congruent, the opposite exterior angles will also be congruent.

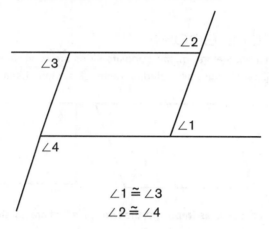

$$\angle 1 \cong \angle 3$$
$$\angle 2 \cong \angle 4$$

Example:

Given the parallelogram below and the exterior angles shown, what is the value of the variable x?

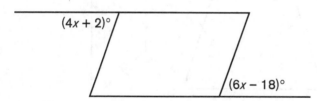

These exterior angles are congruent, because the opposite angles of a parallelogram are congruent. Use algebra:

$6x - 18 = 4x + 2$	Set up an equation.
$6x - 18 - 4x = 4x + 2 - 4x$	Subtract $4x$ from both sides.
$2x - 18 = 2$	Combine like terms.
$2x - 18 + 18 = 2 + 18$	Add 18 to both sides.
$2x = 20$	Combine like terms.
$\dfrac{2x}{2} = \dfrac{20}{2}$	Divide both sides by 2.
$x = 10$	

THE SIDES AND DIAGONALS OF QUADRILATERALS

There are special rules that govern the sides of some quadrilaterals.

 RULE BOOK

In a parallelogram, the opposite sides are congruent. This is true for all parallelograms, including rectangles, rhombuses, and squares.

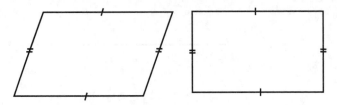

In an isosceles trapezoid, the two sides that are not the bases are congruent.

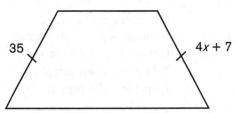

Example:
In the isosceles trapezoid below, what is the value of the variable *x*?

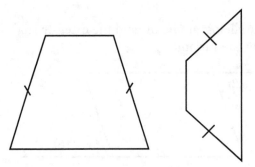

Use algebra:

$4x + 7 = 35$ Set up an equation.
$4x + 7 - 7 = 35 - 7$ Subtract 7 from both sides.
$4x = 28$ Combine like terms.
$\dfrac{4x}{4} = \dfrac{28}{4}$ Divide both sides by 4.
$x = 7$

DIAGONAL of a polygon is any segment that connects two non-consecutive vertices

There are helpful rules that govern the diagonals of some quadrilaterals:

RULE BOOK

The diagonals of every parallelogram bisect each other. This includes rectangles, rhombuses, and squares.

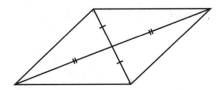

The diagonals of every rectangle are congruent. This includes squares.

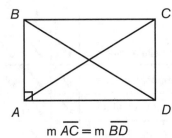

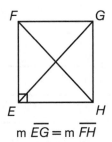

$$m\ \overline{AC} = m\ \overline{BD} \qquad m\ \overline{EG} = m\ \overline{FH}$$

The diagonals of an isosceles trapezoid are congruent.

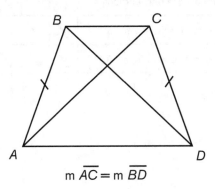

$$m\ \overline{AC} = m\ \overline{BD}$$

RULE BOOK (*continued*)

The diagonals of every rhombus are perpendicular. This includes squares.

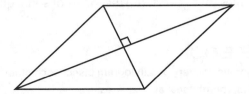

This chart summarizes the facts about quadrilaterals:

	Parallelogram	Rectangle	Rhombus	Square	Isosceles Trapezoid
opposite angles ≅	✓	✓	✓	✓	
opposite sides ≅	✓	✓	✓	✓	
diagonals bisect each other	✓	✓	✓	✓	
diagonals ≅		✓		✓	✓
diagonals are perpendicular			✓	✓	
sum of angle measures is 360°	✓	✓	✓	✓	✓
sum of 2 consecutive angles is 180°	✓	✓	✓	✓	✓

Example:
Given rectangle *LMNO* below, the measure of $\overline{MO}$ = 126. What is the value of the variable *x*?

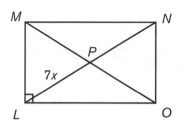

The diagonals of the rectangle are congruent and they also bisect each other. The measure of $\overline{LP}$ is equal to the measure of $\overline{PN}$. Use algebra:

$7x + 7x = 126$	Set up an equation.
$14x = 126$	Combine like terms.
$\dfrac{14x}{14} = \dfrac{126}{14}$	Divide both sides by 14.
$x = 9$	

EXTRA HELP

The website *www.math.com* has an interactive lesson on quadrilaterals. On the left sidebar, click on *Geometry.* Then, under *Polygons,* click on *Quadrilaterals.* There is a series of interactive lessons, followed by a short interactive quiz. Another website is *http://library.thinkquest.org/ 20991/geo/index.html.* Upon reaching this site, click on *Quadrilaterals,* or *Parallelograms,* located on the left-hand list. For more information and practice on working with quadrilaterals, reference the book *Geometry Success in 20 Minutes a Day,* published by LearningExpress.

THE PARTS OF A CIRCLE

Circles are another common plane figure.

CIRCLE the set of all points equidistant from one given point, called the *center*. The center point defines the circle, but is not on the circle.

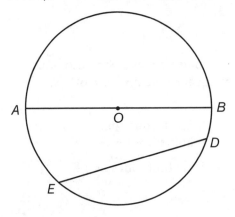

CHORD a line segment whose endpoints are on the circle. In the above figure, there are two chords shown: $\overline{AB}$ and $\overline{DE}$.

DIAMETER of a circle is a chord that passes through the center of the circle. A diameter is shown in the circle above as $\overline{AB}$. The diameter is twice the radius of the circle. This is represented by $d = 2r$. All diameters in a circle are congruent; they have equal measure.

RADIUS of a circle is the line segment whose one endpoint is at the center of the circle and whose other endpoint is on the circle. A radius is shown in the circle above as $\overline{AO}$ or $\overline{OB}$. The radius is one-half the length of the diameter. This is represented by $r = \frac{1}{2}d$. All radii in a circle are congruent; they have equal measure.

TANGENT to a circle is a line that intersects the circle in exactly one point

TANGENT SEGMENT a part of the tangent whose one endpoint is the point on the circle

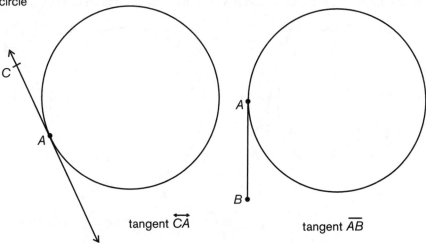

tangent $\overleftrightarrow{CA}$ tangent $\overline{AB}$

SECANT a line that intersects the circle in two points. A secant contains a chord of the circle.

SECANT SEGMENT a segment that extends in one direction beyond the circle, but contains the two points on the circle. One of the points on the circle is the endpoint of the secant segment.

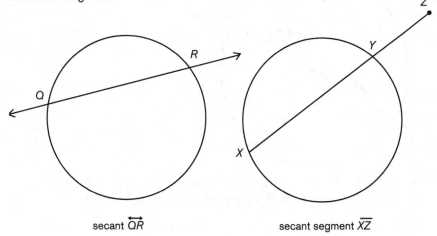

secant $\overleftrightarrow{QR}$ secant segment $\overline{XZ}$

ARC of a circle is a piece of the circle. Arc length can be measured in degrees. The sum of the measures of the arc sections that form the circle adds up to 360°.

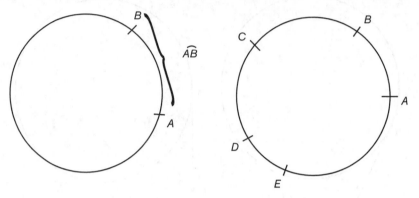

$$m \overparen{AB} + m \overparen{BC} + m \overparen{CD} + m \overparen{DE} + m \overparen{EA} = 360°$$

Example:
In the figure below of circle O, what is the length of $\overline{AB}$?

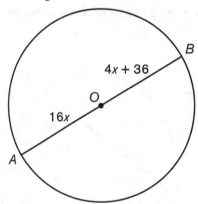

All radii in a circle are congruent. Use algebra to find the value of the variable x. Then, find the length of the diameter, $\overline{AB}$.

$16x = 4x + 36$	Set up an equation.
$16x - 4x = 4x + 36 - 4x$	Subtract $4x$ from both sides.
$12x = 36$	Combine like terms.
$\frac{12x}{12} = \frac{36}{12}$	Divide both sides by 12.
$x = 3$	

Use the value of $x = 3$ to find the length of $\overline{AB}$: $16x + 4x + 36 = 20x + 36 = (20 \times 3) + 36 = 60 + 36 = 96$.

Sometimes, problems are presented in which the angle or arc measures are described as a ratio. In these types of problems, show the measures as the ratio factor times a variable.

Example:
In a given circle, the arc measures *AB*, *BC*, *CD*, *DA* are in the ratio of 2 : 3 : 3 : 4. What is the measure of each of the arcs?

The sum of the degree measure of all of the arcs in a circle is 360°. Assign a variable, like *x*, and express each arc measure as a variable times the ratio factor. Use algebra to solve for *x*, and then find the measure of the arcs.

$2x + 3x + 3x + 4x = 360$	Set up an equation.
$12x = 360$	Combine like terms.
$\dfrac{12x}{12} = \dfrac{360}{12}$	Divide both sides by 12.
$x = 30$	

Now use the value of 30 to find the measure of each of the arcs:
$AB = 2x = 2 \times 30 = 60°$, $BC = CD = 3x = 3 \times 30 = 90°$ each, and
$DA = 4x = 4 \times 30 = 120°$.

ANGLES AND ARCS RELATED TO CIRCLES AND THEIR MEASURE

GLOSSARY

CENTRAL ANGLE an angle whose vertex is the center of the circle. The sum of the measures of the central angles that form the circle adds up to 360°.

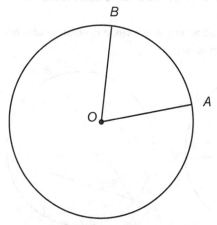

∠*BOA* is a central angle

INSCRIBED ANGLE an angle whose vertex is on the circle, and whose sides pass through two other points on the circle

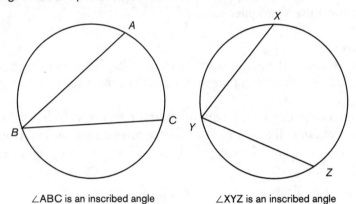

∠ABC is an inscribed angle   ∠XYZ is an inscribed angle

INTERCEPTED ARC an arc contained in the interior of an angle

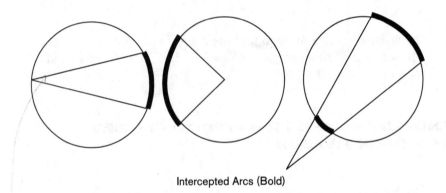

Intercepted Arcs (Bold)

RULE BOOK

The measure of the central angle in a circle is equal to the measure of the intercepted arc.

The measure of an inscribed angle in a circle is equal to one-half the measure of the intercepted arc.

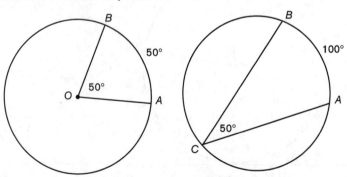

SHORTCUT

A central angle of 180° is a diameter of the circle; it divides the circle in half.

Perpendicular radii form a central angle of 90°; they divide the circle to form a quarter circle.

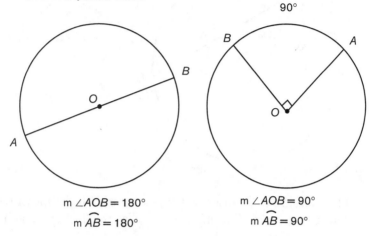

$$m \angle AOB = 180°$$
$$m \overset{\frown}{AB} = 180°$$

$$m \angle AOB = 90°$$
$$m \overset{\frown}{AB} = 90°$$

Example:

Given the following circle, what is the measure of $\overset{\frown}{AB}$?

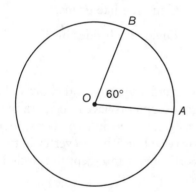

The measure of the intercepted arc is equal to the measure of the central angle that intercepts that arc. $\overset{\frown}{AB} = 60°$ because $\angle BOA = 60°$.

Example:
In the following circle, what is the value of the variable x?

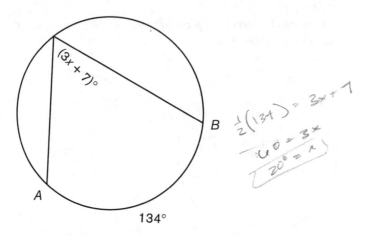

The measure of an inscribed angle is one-half the measure of the arc that it intercepts. The angle is 67°. Use this fact and algebra to find the value of x:

$3x + 7 = 67$	Set up an equation.
$3x + 7 - 7 = 67 - 7$	Subtract 7 from both sides.
$3x = 60$	Combine like terms.
$\dfrac{3x}{3} = \dfrac{60}{3}$	Divide both sides by 3.
$x = 20$	

There are also angles formed by tangents, secants, and chords of a circle. Angles formed by secant and tangent segments have a vertex in the exterior of the circle. These angles intercept the circle in two places; there are two intercepted arcs. Angles formed by two chords have a vertex in the interior of the circle. The angle and its vertical pair intercept the circle in two places; there are two intercepted arcs.

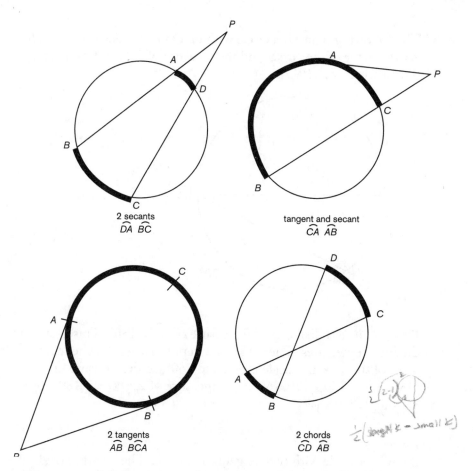

2 secants
$\overset{\frown}{DA}\ \overset{\frown}{BC}$

tangent and secant
$\overset{\frown}{CA}\ \overset{\frown}{AB}$

2 tangents
$\overset{\frown}{AB}\ \overset{\frown}{BCA}$

2 chords
$\overset{\frown}{CD}\ \overset{\frown}{AB}$

There are rules that govern the measure of these angles.

RULE BOOK

Two secant segments, two tangent segments, or a tangent and a secant form an angle exterior to the circle. The measure of this angle is one-half of the difference between the intercepted arcs of the angle.

$$m \angle APB = \frac{1}{2}(y - x)$$

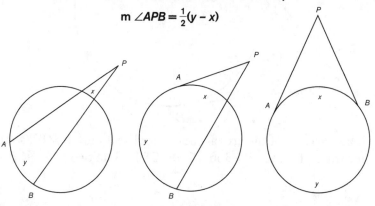

Example:
The following figure is a circle with tangent and secant shown. $\overline{AC}$ is a diameter of the circle, and the measure of $\overparen{CB}$ = 120°. What is the measure of ∠*APC*?

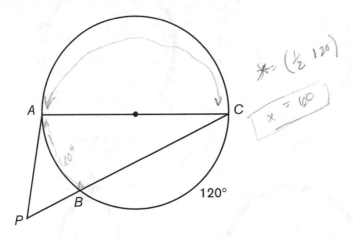

Because $\overline{AC}$ is a diameter, the measure of $\overparen{AC}$ = 180°. This is one of the intercepted arcs. The other intercepted arc is $\overparen{AB}$. The measure of all of the arcs in a circle is equal to 360°, so the measure of $\overparen{AB}$ = 360 – 180 – 120 = 60°. The measure of ∠*APC* = $\frac{1}{2}$(180 – 60) = $\frac{1}{2}$(120) = $\frac{120}{2}$ = 60°.

Example:
Below is a circle with two tangents as shown. The measure of $\overparen{AB}$ is 110°. What is the measure of ∠*APB*?

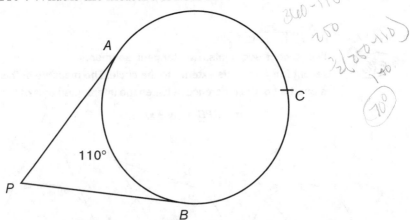

The measure of the arcs around a circle sum up to 360°, so the measure of arc *ACB* = 360 – 110 = 250°. The two arcs, *ACB* and *AB*, are the intercepted arcs. Use the formula to calculate the measure of ∠APB:

measure of ∠APB = $\frac{1}{2}(250 - 110) = \frac{1}{2}(140) = \frac{140}{2} = 70°$

 RULE BOOK

A pair of vertical angles is formed by the intersection of two chords of a circle. The measure of either of these angles is equal to one-half of the sum of the intercepted arcs of the angles.

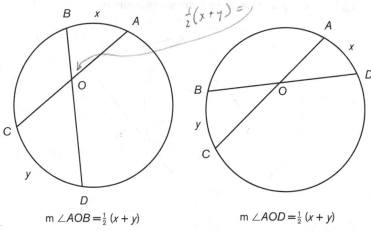

$$m \angle AOB = \frac{1}{2}(x + y)$$

$$m \angle AOD = \frac{1}{2}(x + y)$$

Example:
Using the following circle, determine the measure of ∠AOD.

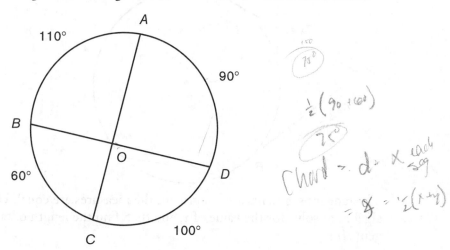

The two intercepted arcs are $\overset{\frown}{AD}$ and $\overset{\frown}{BC}$. Use the formula to calculate the angle measure. The measure of ∠AOD = $\frac{1}{2}(90 + 60)$ = $\frac{1}{2}(150) = \frac{150}{2} = 75°$.

SEGMENT MEASURES AS RELATED TO CIRCLES

Just as there are rules concerning the angles related to the circle, there are also rules that govern the segments formed by secant and tangent segments and chords.

 RULE BOOK

Two tangent segments to a circle that share a common endpoint are congruent; they have the same measure.

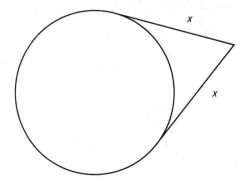

Example:
In the following graphic of a circle with two tangents shown, what is the length of tangent $\overline{AP}$?

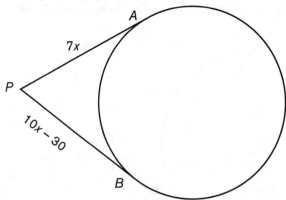

The tangent segments are congruent; the measures are equal. Use algebra to solve for the value of x, and then find the length of tangent $\overline{AP}$:

$10x - 30 = 7x$	Set up an equation.
$10x - 30 - 7x = 7x - 7x$	Subtract $7x$ from both sides.
$3x - 30 = 0$	Combine like terms.

$3x - 30 + 30 = 0 + 30$ Add 30 to both sides.

$3x = 30$ Combine like terms.

$\dfrac{3x}{3} = \dfrac{30}{3}$ Divide both sides by 3.

$x = 10$

Use this value of 10 to find the length of $\overline{AP} = 7x = 7 \times 10 = 70$.

A secant to a circle has a portion of its segment in the interior of the circle (a chord of the circle), and a portion in the exterior of the circle, which will be called the outer piece.

RULE BOOK

When two secants intersect in the exterior of a circle, the product of the measure of the outer piece and the entire secant segment is equal to the product of the corresponding pieces of the other segment. This is easier to show with a diagram and labels.

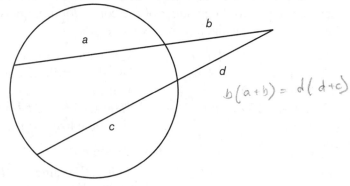

$$b(a + b) = d(c + d)$$

When a secant and a tangent segment intersect in the exterior of a circle, the product of the measure of the outer piece and the entire secant segment is equal to the square of the tangent segment.

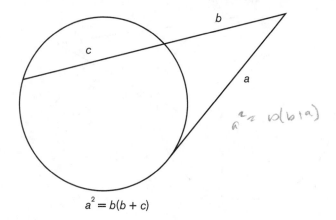

$$a^2 = b(b + c)$$

Example:

Given the following circle with secants shown, what is the value of chord $\overline{AB}$?

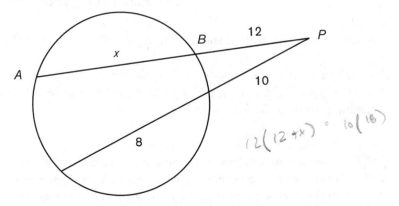

The product of the outer piece times the whole segment of one secant is equal to the product of the corresponding pieces of the other secant. Use algebra to solve for x:

$12(x + 12) = (18 \times 10)$	Set up an equation.
$12x + (12 \times 12) = (18 \times 10)$	Use the distributive property.
$12x + 144 = 180$	Multiply.
$12x + 144 - 144 = 180 - 144$	Subtract 144 from both sides.
$12x = 36$	Combine like terms.
$\dfrac{12x}{12} = \dfrac{36}{12}$	Divide both sides by 12.
$x = 3$	This is the value of $\overline{AB}$.

Example:

In the following circle with tangent and secant shown, what is the length of $\overline{AP}$?

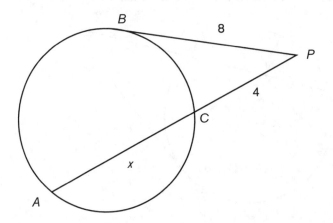

The square of the tangent measure is equal to the product of the outer piece and the whole secant segment. Use algebra to find the value of the variable x, and then use this value to find the length of $\overline{AP}$:

$4(x + 4) = 8^2$	Set up an equation.
$4x + (4 \times 4) = 8^2$	Use the distributive property.
$4x + 16 = 64$	Multiply.
$4x + 16 - 16 = 64 - 16$	Subtract 16 from both sides.
$4x = 48$	Combine like terms.
$\dfrac{4x}{4} = \dfrac{48}{4}$	Divide both sides by 4
$x = 12$	

Use this value of 12 to find the length of $\overline{AP} = 12 + 4 = 16$.

CALCULATOR TIPS

Locate the "square" key and the square root key on your calculator. They will be helpful when working with the tangent segment measurement problems. The keys will look similar to the following graphic. Often, the square root function is the second function that corresponds to the "square" key.

RULE BOOK

When two chords of a circle intersect, the product of the measures of the two segments of one chord equals the product of the measures of the segments of the other chord.

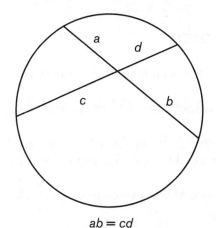

$$ab = cd$$

Example:
In the circle below, what is the length of $\overline{AO}$?

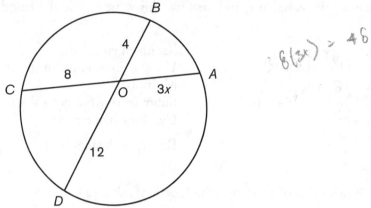

$6(3x) = 48$

For the two intersecting chords, the product of the measures of the two segments of one chord equals the product of the measures of the two segments of the other chord. Use algebra and solve for x; find the length of $\overline{AO}$:

$3x \times 8 = 12 \times 4$ Set up an equation.
$24x = 48$ Multiply.
$\dfrac{24x}{24} = \dfrac{48}{24}$ Divide both sides by 24.
$x = 2$

Use this value of 2 to find the length of $\overline{AO} = 3x = 3 \times 2 = 6$.

 EXTRA HELP

The website *http://library.thinkquest.org/20991/geo/index.html* has a lesson on circle relationships. Upon reaching this site, click on *Circles*, located on the left-hand list. Follow the tutorial and take the quiz.

TIPS AND STRATEGIES

When solving problems related to quadrilaterals and circles, remember:

- Quadrilaterals are classified according to the Venn diagram shown in this chapter.
- The sum of the measure of the angles in a quadrilateral is 360°.
- The sum of the measure of the exterior angles in a quadrilateral is 360°.
- The sum of the measure of any two consecutive angles in a parallelogram is 180°.

- Opposite sides and opposite angles in a parallelogram are congruent.
- In an isosceles trapezoid, the base angles and the sides that are not the bases are congruent.
- Know the facts about the diagonals of quadrilaterals as shown in the chart in this chapter.
- Be familiar with circle vocabulary.
- The measure of a central angle in a circle is equal to the measure of the intercepted arc.
- The measure of an inscribed angle in a circle is equal to one-half of the measure of the intercepted arc.
- Be familiar with the rules for finding other angles and segment lengths dealing with chords, secants, and tangents of a circle.

PRACTICE QUIZ

Try these 25 problems. When you are finished, review the answers and explanations to see if you have mastered this concept.

1. Which of the following statements is ALWAYS true?
   a. The diagonals of all parallelograms are congruent.
   b. All rectangles are squares.
   c. All trapezoids have two congruent sides.
   d. All squares are rhombuses.
   e. All parallelograms have a 90° angle.

2. In the following parallelogram *ABCD*, what is the measure of ∠*ADC*?

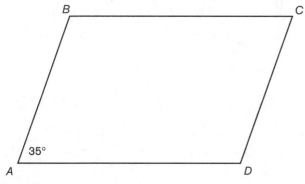

   a. 145°
   b. 65°
   c. 325°
   d. 165°
   e. 45°

3. Given rectangle $WXYZ$ below, what is the length of $\overline{WZ}$?

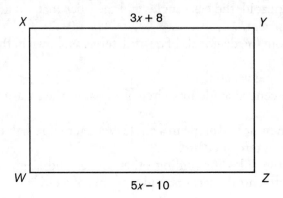

a. 11
b. 9
c. 35
d. 30
e. 70

4. In square $ABCD$, if $\overline{BD} = 3x + 17$ and $\overline{AC} = 7x - 23$, what is the value of the variable x?

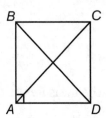

a. 10

b. 47

c. 7

d. $\frac{3}{2}$

e. 38

5. Given isosceles trapezoid *PQRS*, what is the measure of ∠*QPS*?

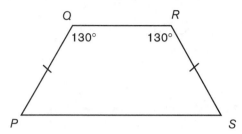

a. 100°
b. 70°
c. 45°
d. 20°
e. 50°

6. In parallelogram WXYZ, what is the length of $\overline{WC}$?
$\overline{XC} = 20$, $\overline{WY} = 10x$, and $\overline{CZ} = 2x + 8$

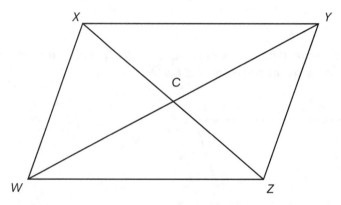

a. 20
b. 30
c. 50
d. 15
e. 60

7. Which statement below is FALSE?
a. A rhombus has four congruent sides.
b. A rhombus is a parallelogram.
c. The diagonals of a rhombus bisect each other.
d. The diagonals of a rhombus are perpendicular.
e. All rhombuses have 90° angles.

8. Given trapezoid *GHIJ*, what is the value of the variable *x*?

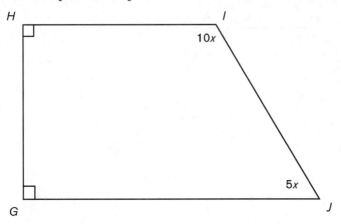

a. 24°
b. 75°
c. 60°
d. 12°
e. 120°

9. The following figure is parallelogram *ABCD* with exterior angles shown. What is the measure of ∠*BAD*?

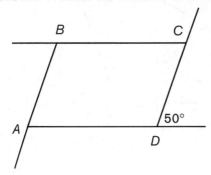

a. 50°
b. 150°
c. 130°
d. 30°
e. 80°

10. What is the value of the variable x in the following rhombus $RSTU$?

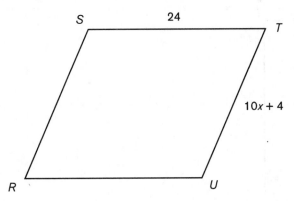

a. 20
b. 2
c. 2.8
d. 4
e. 4.4

11. In isosceles trapezoid $LMNO$, what is the length of $\overline{MN}$?

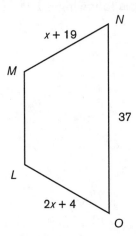

a. 37
b. 18
c. 16.5
d. 15
e. 34

12. Which of the following statements is true for rectangle $JKLM$?

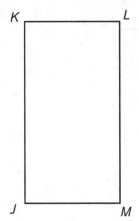

a. $\overline{JL} \cong \overline{KM}$
b. $\overline{KM}$ bisects $\overline{JL}$.
c. $\overline{LM} \cong \overline{KJ}$
d. Choices **a**, **b**, and **c** are all true.
e. None of the above is true.

13. Use the diagram to determine which of the following is FALSE.

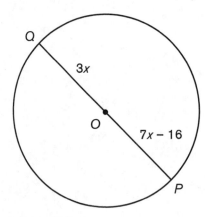

a. $\overline{OQ} \cong \overline{OP}$
b. $x = 4$
c. $\overline{QP} = 12$
d. $\overline{QP} = 10x - 16$
e. $\overline{OP} = 12$

14. What is the measure of ∠*MOP*?

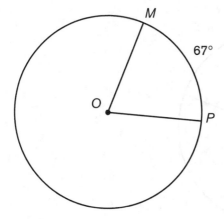

   a. 34.5°
   b. 17°
   c. 67°
   d. 113°
   e. 134°

15. What is the measure of ∠*BAC*?

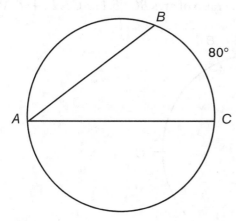

   a. 80°
   b. 40°
   c. 100°
   d. 10°
   e. 20°

16. Given that $\overline{WY}$ is a diameter of the following circle, what is the measure of $\widehat{WX}$?

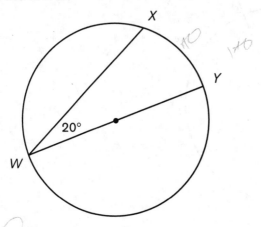

a. 140°
b. 180°
c. 20°
d. 160°
e. 200°

17. In the following circle, the ratio of arcs $BC : BA : AC$ is $2 : 4 : 6$. What is the measure of $\widehat{BA}$?

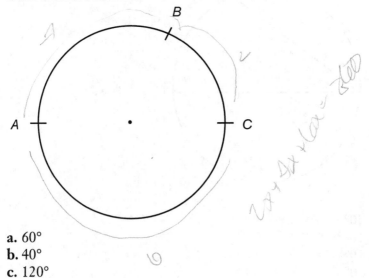

a. 60°
b. 40°
c. 120°
d. 240°
e. 20°

18. Use the following circle to find the measure of ∠AOD.

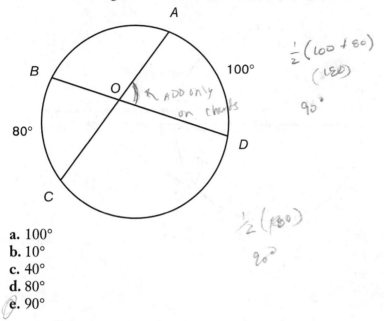

a. 100°
b. 10°
c. 40°
d. 80°
e. 90°

19. Use the following circle with tangents shown to find the measure of ∠APB.

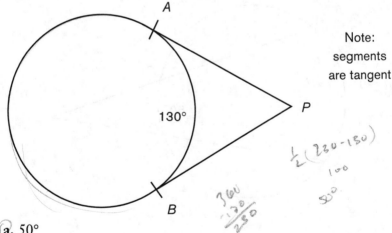

Note: segments are tangent

a. 50°
b. 120°
c. 240°
d. 130°
e. 65°

20. Given that $\overline{AB}$ is a diameter of circle O below, what is the measure of $\angle BPC$?

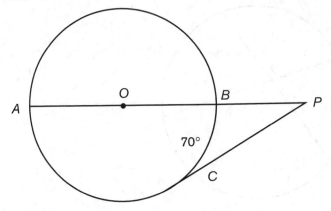

a. 70°
b. 20°
c. 80°
d. 60°
e. 35°

21. In the following circle with two secants shown, what is the measure of $\angle BPC$?

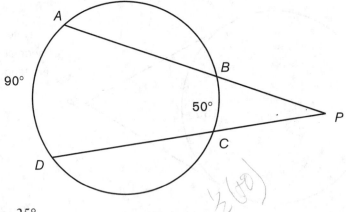

a. 25°
b. 20°
c. 45°
d. 50°
e. 10°

22. What is the value of the variable *x* shown below in the circle with two tangents?

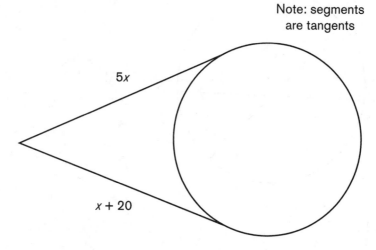

Note: segments
are tangents

5*x*

x + 20

a. 25
b. 12.5
c. 625
d. 5
e. 50

23. What is the length of secant $\overline{AP}$ in the following circle?

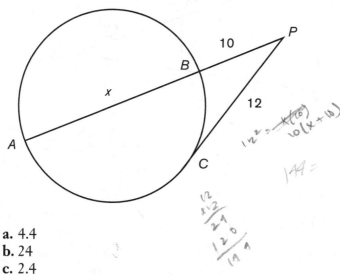

a. 4.4
b. 24
c. 2.4
d. 2
e. 14.4

24. In the circle with two chords shown below, what is the length of $\overline{OB}$?

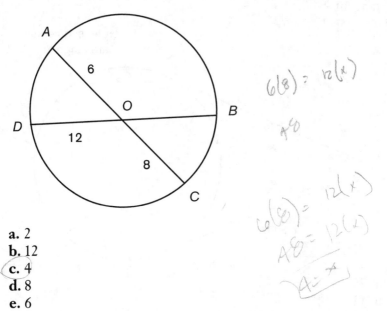

$6(8) = 12(x)$

A^{-0}

$6(8) = 12(x)$

$6(8) = 12(x)$

$48 = 12(x)$

$4 = x$

a. 2
b. 12
c. 4
d. 8
e. 6

25. In the circle with two secants shown below, what is the length of $\overline{AP}$?

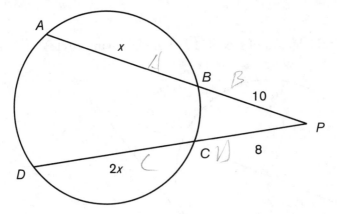

a. 20
b. 16
c. 10
d. 6
e. 12

ANSWERS

Here are the answers, with detailed explanations, for all the practice problems. Study the explanations for any problems that you answered incorrectly. Then, go back and try those problems again. Refer back to the chapter for further study if needed.

1. d. All squares are rhombuses; a rhombus has four congruent sides. If the diagonals of a parallelogram are congruent, then it is a rectangle; this is not true for every parallelogram. Only some rectangles are squares. Some trapezoids, but not all, have two congruent sides. Rectangles are the only type of parallelogram that have 90° angles.

2. a. The measure of any two consecutive angles in a parallelogram add up to 180°. The measure of the angle is $180 - 35 = 145°$.

3. c. Opposite sides of a rectangle are congruent. Use algebra to find the value of the variable x, then find the length of $\overline{WZ}$:

$3x + 8 = 5x - 10$	Set up an equation.
$3x + 8 - 3x = 5x - 10 - 3x$	Subtract $3x$ from both sides.
$8 = 2x - 10$	Combine like terms.
$8 + 10 = 2x - 10 + 10$	Add 10 to both sides.
$18 = 2x$	Combine like terms.
$\frac{18}{2} = \frac{2x}{2}$	Divide both sides by 2.
$9 = x$	

$\overline{WZ} = 5x - 10 = (5 \times 9) - 10 = 45 - 10 = 35$

4. a. The diagonals of a square are congruent; this is true for any rectangle. Use algebra to find the value of x:

$3x + 17 = 7x - 23$	Set up an equation.
$3x + 17 - 3x = 7x - 23 - 3x$	Subtract $3x$ from both sides.
$17 = 4x - 23$	Combine like terms.
$17 + 23 = 4x - 23 + 23$	Add 23 to both sides.
$40 = 4x$	Combine like terms.
$\frac{40}{4} = \frac{4x}{4}$	Divide both sides by 4.
$10 = x$	

5. e. In an isosceles trapezoid, the base angles are congruent. In addition, the sum of the measure of the angles in a trapezoid is 360°. The measure of the sum of the base angles is $360 - 130 - 130 = 100°$. The measure of $\angle QPS$ is one-half of 100°, which is 50°.

6. b. The diagonals of a parallelogram bisect each other. The two smaller segments of each diagonal are therefore congruent. Use this fact to find the value of the variable x:

$2x + 8 = 20$	Set up an equation.
$2x + 8 - 8 = 20 - 8$	Subtract 8 from both sides.
$2x = 12$	Combine like terms.
$\dfrac{2x}{2} = \dfrac{12}{2}$	Divide both sides by 2.
$x = 6$	

Use this value to find the length of $\overline{WY}$, $10x = 10 \times 6 = 60$. The length of $\overline{WC}$ is one-half of the length of $\overline{WY}$, so it is 30.

7. e. Only some rhombuses have 90° angles, that is, those that are squares. Statements **a** through **d** are always true of every rhombus.

8. d. The sum of the measure of the angles in a trapezoid is 360°. There are two 90° angles, as shown by the small box written in the angles. Use algebra to solve for x:

$10x + 5x + 90 + 90 = 360$	Set up an equation.
$15x + 180 = 360$	Combine like terms.
$15x + 180 - 180 = 360 - 180$	Subtract 180 from both sides.
$15x = 180$	Combine like terms.
$\dfrac{15x}{15} = \dfrac{180}{15}$	Divide both sides by 15.
$x = 12$	

9. a. Any exterior angle and its adjacent interior angle form a linear pair. From the figure, the measure of $\angle ADC = 180 - 50 = 130°$. Any two consecutive interior angles of a parallelogram add up to 180. The measure of $\angle BAD = 180 - 130 = 50°$.

10. b. All of the sides of a rhombus are congruent. Use algebra to solve for x:

$10x + 4 = 24$	Set up an equation.
$10x + 4 - 4 = 24 - 4$	Subtract 4 from both sides.
$10x = 20$	Combine like terms.
$\dfrac{10x}{10} = \dfrac{20}{10}$	Divide both sides by 10.
$x = 2$	

11. e. In an isosceles trapezoid, the sides that are not bases are congruent. These are the sides marked as expressions. Use algebra to
solve for x, and then find the length of $\overline{MN}$.

$2x + 4 = x + 19$	Set up an equation.
$2x + 4 - x = x + 19 - x$	Subtract x from both sides.
$x + 4 = 19$	Combine like terms.
$x + 4 - 4 = 19 - 4$	Subtract 4 from both sides
$x = 15$	

Use the value of 15 to find the length of $\overline{MN} = x + 19 = 15 + 19$
$= 34$.

12. d. Choices **a**, **b**, and **c** are true. The diagonals of a rectangle are congruent. The diagonals of a rectangle bisect each other. Opposite
sides of a rectangle are congruent.

13. c. Use the fact that all radii in a circle are congruent. Statement **a** is
therefore true. After finding the value of x, determine which of
the other statements is false.

$7x - 16 = 3x$	Set up an equation.
$7x - 16 - 3x = 3x - 3x$	Subtract $3x$ from both sides.
$4x - 16 = 0$	Combine like terms.
$4x - 16 + 16 = 0 + 16$	Add 16 to both sides.
$4x = 16$	Combine like terms.
$\dfrac{4x}{4} = \dfrac{16}{4}$	Divide both sides by 4.
$x = 4$	

$\overline{QP} = 3x + 7x - 16 = 10x - 16$; $\overline{QP} = (10 \times 4) - 16 = 40 - 16 = 24$.
Statement **c** is false.

14. c. The measure of a central angle is equal to the measure of the arc
it intercepts.

15. b. The measure of an inscribed angle is equal to one-half the measure of the arc it intercepts. The measure of $\angle BAC = 80 \div 2 = 40°$.

16. a. The measure of the arc that intercepts a diameter is 180°. The
measure of $\overset{\frown}{WY}$ is 180°. The measure of $\overset{\frown}{XY}$ is 40°; it is twice the
measure of the inscribed angle that it intercepts. The measure of
$\overset{\frown}{WX} = WY - XY$, or $180 - 40 = 140°$.

17. c. The sum of the measures of the three arcs in the circle is equal to the degree measure of the circle, which is 360°. Use algebra to assign the arc measures as $2x$, $4x$, and $6x$, respectively, and then solve for x. Find the measure of $\overset{\frown}{BA}$.

$2x + 4x + 6x = 360$	Set up an equation.
$12x = 360$	Combine like terms.
$\dfrac{12x}{12} = \dfrac{360}{12}$	Divide both sides by 12.
$x = 30$	

Use the value of 30 to find the measure of arc $BA = 4x = 4 \times 30 = 120°$.

18. e. The measure of an interior angle formed by the intersection of two chords is one-half of the sum of the measure of the two arcs that the chords intercept. The measure of $\angle AOD = 100 + \dfrac{80}{2} = \dfrac{180}{2} = 90°$.

19. a. The measure of the exterior angle formed by two tangents is one-half of the difference between the outer arc and the inner arc intercepted by the two tangents. The outer arc is the degree measure of the circle minus the measure of the inner arc; $360 - 130 = 230°$. The measure of $\angle APB = 230 - \dfrac{130}{2} = \dfrac{100}{2} = 50°$.

20. b. The measure of the exterior angle formed by a tangent and a secant is one-half of the difference between the outer arc and the inner arc intercepted by these segments. Because $\overline{AB}$ is a diameter, $\overset{\frown}{AB} = 180°$; the measure of $\overset{\frown}{AC} = 180 - 70 = 110°$. The measure of $\angle BPC = 110 - \dfrac{70}{2} = \dfrac{40}{2} = 20°$.

21. b. The measure of the exterior angle formed by two tangents is one-half of the difference between the outer arc and the inner arc intercepted by the two tangents. The measure of $\angle BPC = 90 - \dfrac{50}{2} = \dfrac{40}{2} = 20°$.

22. d. The lengths of tangents to a circle are congruent. Use algebra to solve for x:

$5x = x + 20$	Set up an equation.
$5x - x = x + 20 - x$	Subtract x from both sides.
$4x = 20$	Combine like terms.
$\dfrac{4x}{4} = \dfrac{20}{4}$	Divide both sides by 4.
$x = 5$	

23. e. The measure of a secant and a tangent to a circle are related by the fact that the square of the tangent measure is equal to the product of the outer piece and the whole piece of the secant. Use algebra to solve for x:

$10(x + 10) = 12^2$	Set up an equation.
$10x + (10 \times 10) = 12^2$	Use the distributive property.
$10x + 100 = 144$	Multiply.
$10x + 100 - 100 = 144 - 100$	Subtract 100 from both sides.
$10x = 44$	
$\dfrac{10x}{10} = \dfrac{44}{10}$	Divide both sides by 10.
$x = 4.4$	

The length of $\overline{AP} = x + 10 = 4.4 + 10 = 14.4$.

24. c. Multiply the two segments of one chord together. This value is equal to the product of the two segments of the other chord shown. Use algebra and assign the missing segment length to the variable x. Solve for x, and this will be the length of $\overline{OB}$.

$12x = 8 \times 6$	Set up an equation.
$12x = 48$	Multiply.
$\dfrac{12x}{12} = \dfrac{48}{12}$	Divide both sides by 12.
$x = 4$	This is the length of $\overline{OB}$.

25. b. For the two secants shown, the lengths are in the relationship of (outer piece) times (whole segment) = (outer piece) times (whole segment). Use algebra to find the value of x, and then find the length of $\overline{AP}$.

$10(x + 10) = 8(2x + 8)$	Set up an equation.
$10x + (10 \times 10) = 16x + (8 \times 8)$	Use the distributive property.
$10x + 100 = 16x + 64$	Multiply.
$10x + 100 - 10x = 16x + 64 - 10x$	Subtract $10x$ from both sides.
$100 = 6x + 64$	Combine like terms.
$100 - 64 = 6x + 64 - 64$	Subtract 64 from both sides.
$36 = 6x$	Combine like terms.
$\dfrac{36}{6} = \dfrac{6x}{6}$	Divide both sides by 6.
$6 = x$	

Use the value of 6 to find the length of $\overline{AP} = x + 10 = 6 + 10 = 16$.

Perimeter and Area

The word *geometry* means "measure of the earth." We constantly measure things in real life. We measure for fences and carpets; we measure fabric and wallpaper. A basic understanding of geometry is crucial to success on your upcoming test. Take this short benchmark quiz to see how much of your studies should be concentrated on this chapter.

BENCHMARK QUIZ

1. Find the area.

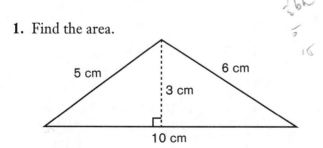

    a. 15 cm^2
    b. 25 cm^2
    c. 60 cm^2
    d. 50 cm^2
    e. 30 cm^2

2. Find the perimeter.

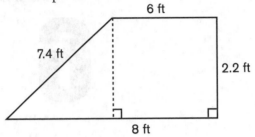

a. 10.2 ft
b. 23.6 ft
c. 20.4 ft
d. 43.2 ft
e. 15 ft

3. Find the area.

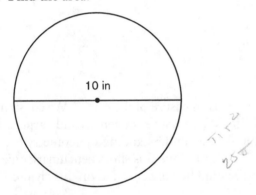

a. 5π in^2
b. 100π in^2
c. 25π in^2
d. 100 in^2
e. 25 in^2

4. Find the area.

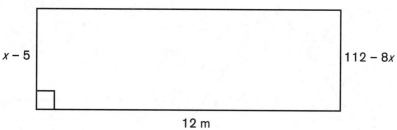

$x - 5$ | 12 m | $112 - 8x$

a. 96 m²
b. 60 m²
c. 120 m²
d. 80 m²
e. 960 m²

$x - 5 = 112 - 8x$

$9x = 117$

$x \sim 13$

$12 \cdot \frac{98}{8}$

96

5. Find the circumference.

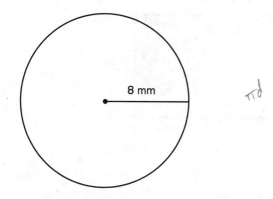

8 mm

πd

a. 16 mm
b. 8π mm
c. 64π mm
d. 256π mm
e. 16π mm

6. Find the perimeter of the isosceles triangle.

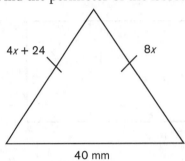

4x + 24   8x

40 mm

   a. 72 mm
   b. 320 mm
   c. 40 mm
   d. 136 mm
   e. 120 mm

7. Find the area of the shaded region.

book
is
wrong

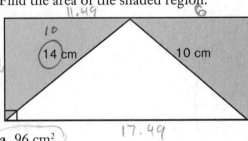

11.49   6

10

14 cm   10 cm   8 cm

17.49

   a. 96 cm²
   b. 48 cm²
   c. 120 cm²
   d. 24 cm²
   e. 144 cm²

8. Find the area.

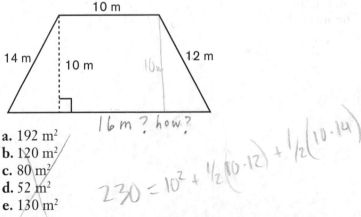

10 m

14 m   10 m   10m   12 m

16 m ? how?

   a. 192 m²
   b. 120 m²
   c. 80 m²
   d. 52 m²
   e. 130 m²

$$230 = 10^2 + \tfrac{1}{2}(10 \cdot 12) + \tfrac{1}{2}(10 \cdot 14)$$

9. Find the area of the irregular figure. Use 3.14 for π.

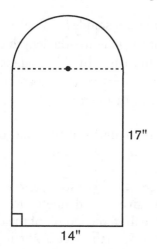

17"

14"

a. 391.86 in²
b. 853.44 in²
c. 545.72 in²
d. 314.93 in²
e. 281.96 in²

10. Find the area of the shaded region. Use 3.14 for π.

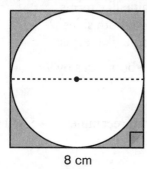

8 cm

a. 64 cm²
b. 114.24 cm²
c. 13.76 cm²
d. 137.6 cm²
e. 50.24 cm²

BENCHMARK QUIZ SOLUTIONS

1. a. The area of a triangle is $A = \frac{1}{2}bh$, where A stands for area, b stands for the base length, and h stands for the height—the length of the segment that is perpendicular to the base. In this triangle, the base is 10 cm and the height is 3 cm. $A = \frac{1}{2} \times 10 \times 3$ and multiplying them all together yields an area of 15 cm².

2. b. To find perimeter, you add up all the lengths of the sides of the polygon. The perimeter is 7.4 + 6 + 2.2 + 8 = 23.6 ft.

3. c. The area of a circle is $A = \pi r^2$, where π is a constant, and r is the radius of the circle. The problem gives the diameter to be 10 inches. The radius, 5 inches, is one-half of the length of the diameter. Using the formula, $A = \pi \times 5 \times 5$, so the area is 25π in².

4. a. Opposite sides of a rectangle are congruent. Use this fact and algebra to solve for the variable x. Then, find the length of the side of the rectangle. Multiply this by the side of length 12 to get the area:

$x - 5 = 112 - 8x$	Set up an equation.
$x - 5 + 8x = 112 - 8x + 8x$	Add $8x$ to both sides.
$9x - 5 = 112$	Combine like terms.
$9x - 5 + 5 = 112 + 5$	Add 5 to both sides.
$9x = 117$	Combine like terms.
$\frac{9x}{9} = \frac{117}{9}$	Divide both sides by 9.
$x = 13$	

Use this value to find the side of the rectangle: $x - 5 = 13 - 5 = 8$ m. The area is $8 \times 12 = 96$ m².

5. e. Circumference is found by the formula $C = \pi d$, where C is the circumference, π is a constant and d is the diameter. The radius of 8 mm is given, and diameter is twice the radius. The diameter is 16 mm; $C = 16\pi$ mm.

6. d. This is an isosceles triangle; the two sides marked as algebraic expressions are congruent. Use algebra to find the value of x. Then, find the length of these congruent sides. Add up all the sides to find the perimeter.

$8x = 4x + 24$	Set up an equation.
$8x - 4x = 4x + 24 - 4x$	Subtract $4x$ from both sides.

$4x = 24$ Combine like terms.

$\frac{4x}{4} = \frac{24}{4}$ Divide both sides by 4.

$x = 6$

Use this value to find the length of the congruent sides: $8x = 8 \times 6 = 48$ mm. The perimeter is $48 + 48 + 40 = 136$ mm.

7. b. The area of the shaded region is the area of the outer figure, a rectangle, minus the area of the inner figure, a triangle; $A = A_{rectangle} - A_{triangle}$. Using the area formulas, you get $A = bh - \frac{1}{2}bh$; $A = (12 \times 8) - \frac{1}{2}(12 \times 8) = 96 - 48 = 48$ cm^2.

8. e. The area of a trapezoid is $A = \frac{1}{2}h(b_1 + b_2)$, where A stands for area, b_1 and b_2 are the lengths of the parallel bases, and h is the height—the length of the segment perpendicular to the bases. In this trapezoid, the height = 10 m, since it is perpendicular to the bases. The bases are the parallel sides, 16 m and 10 m. Substitute the given information into the formula: $A = \frac{1}{2} \times 10 \times (16 + 10)$ to get $A = \frac{1}{2} \times 10 \times (26)$. Multiply all terms on the right together to yield an area of 130 m^2.

9. d. This figure is a rectangle and one-half of a circle. The area will be $A = A_{rectangle} + (A_{circle}$ divided by 2); $A = bh + \pi r^2 \div 2$. The radius is one-half of the width of the rectangle; the radius is 7; $A = (14 \times 17) + (3.14 \times 7^2) \div 2 = 238 + 76.93 = 314.93$ in^2.

10. c. The area of the shaded region is found by taking the area of the outer figure, the square, and subtracting out the area of the inner figure, the circle. $A = A_{square} - A_{circle}$. Using the area formulas, you get $A = s^2 - \pi r^2$. Notice that the diameter of the circle is the same length as the side of the square. So, the radius is one-half of 8. The radius is 4 cm. Substitute in the given lengths and value of π to get $A = 8 \times 8 - 3.14 \times 4 \times 4$. Using the order of operations, you multiply first, from left to right; $A = 64 - 50.24$, or $A = 13.76$ cm^2.

BENCHMARK QUIZ RESULTS

If you answered 8–10 questions correctly, you have a good understanding of area and perimeter. Perhaps the questions you answered incorrectly deal with one specific area in this chapter. Read over the chapter, concentrating on those areas of weakness. Proceed to the Practice Quiz to try to improve your score.

If you answered 4–7 questions correctly, there are several areas you need to review. Carefully read through the lesson in this chapter for review and skill-building. Work carefully through the examples and pay attention to the sidebars that refer you to definitions, hints, and shortcuts. Get additional practice on geometry by visiting the suggested websites and taking the quiz at the end of the chapter.

If you answered 1–3 questions correctly, you need to spend some time on the topics in this chapter. First, carefully read this chapter and concentrate on the sidebars and visual aids that will help with comprehension. Go to the suggested websites in the Extra Help sidebar in this chapter, which will help with understanding and will provide extended practice. You may also want to refer to *Geometry Success in 20 Minutes a Day*, Lessons 12, 13, and 16, published by LearningExpress.

JUST IN TIME LESSON—PERIMETER AND AREA

There are many applications of geometry in mathematical problems. This chapter will describe ways that geometry is used for measurement.

The topics in this chapter are:

- Perimeter of Polygons
- Area of Common Polygons
- Perimeter and Area of a Circle
- Area of Irregular Shaped Figures
- Area of Shaded Regions

PERIMETER OF POLYGONS

Perimeter is the measure AROUND a polygon. Perimeter is an addition concept; it is a linear, one-dimensional measurement.

 RULE BOOK

To find the perimeter of a polygon, add up all of the lengths of the sides of the figure. Be sure to name the units.

Example:
Find the perimeter.

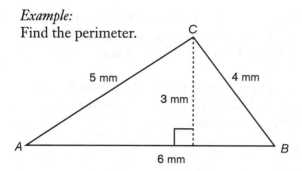

Add up the lengths of all sides, that is *AB*, *BC*, and *AC*. The height, 3 mm, is information that is not needed to calculate the perimeter. Substitute in to get $6 + 4 + 5 = 15$ mm.

SHORTCUT

In a rectangle, like all parallelograms, the opposite sides are parallel and congruent. Perimeter can be found using the formula $P = 2l + 2w$, where *P* is the perimeter, *l* is the length, and *w* is the width.

For a square, or any rhombus, the perimeter can be found by $P = 4s$, where *P* is the perimeter and *s* is the length of one of the sides.

Example:
Find the perimeter of the rectangle.

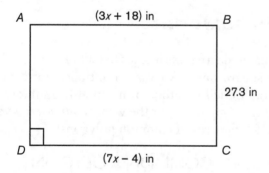

In a rectangle, like all parallelograms, the opposite sides are parallel and congruent. In this example, $\overline{AB} \cong \overline{DC}$. Use algebra to solve for the variable *x*.

$7x - 4 = 3x + 18$	Set up an equation.
$7x - 4 - 3x = 3x + 18 - 3x$	Subtract $3x$ from both sides.
$4x - 4 = 18$	Combine like terms.
$4x - 4 + 4 = 18 + 4$	Add 4 to both sides.
$4x = 22$	Combine like terms.
$\frac{4x}{4} = \frac{22}{4}$	Divide both sides by 4.
$x = 5.5$	

Use this value to find the length of $\overline{AB}$ and $\overline{DC}$; $3x + 18 = (3 \times 5.5)$ $+ 18 = 34.5$. Use the shortcut for the perimeter of a rectangle: $P =$ $(2 \times 34.5) + (2 \times 27.3) = 69 + 54.6 = 123.6$ inches.

Be alert when working with geometry problems to make sure that the units are consistent. If they are different, a conversion must be made before calculating perimeter or area.

Example:
Find the perimeter of the trapezoid.

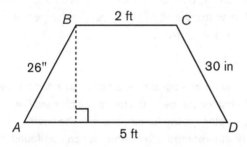

Change the lengths of the sides given in feet to be inches; 1 ft = 12 in; $BC = 2$ ft $= 24$ in, and $AD = 5$ ft $= 60$ in. Now add up the four sides: $24 + 60 + 30 + 26 = 140$ in.

AREA OF COMMON POLYGONS

Area is a measure of how many square units it takes to COVER a closed figure. Area is measured in square units. Area is a multiplication concept, where two measures are multiplied together. You can also think of units being multiplied together: cm $\times$ cm $= cm^2$, or the words *centimeters squared*. There are formulas to use for the area of common polygons:

RULE BOOK—COMMON POLYGON FORMULAS

A stands for area, *b* stands for base, *h* stands for height (which is perpendicular to the base), and b_1 and b_2 are the parallel sides of a trapezoid.

Area of a triangle	$A = \frac{1}{2}bh$
Area of a parallelogram	$A = bh$
Area of a trapezoid	$A = \frac{1}{2}h(b_1 + b_2)$

Be sure to include square units in your answer.

Recall that rectangles, squares, and rhombuses are parallelograms. For rectangles and squares, the height is also a side of the polygon.

Example:
Find the area of the triangle.

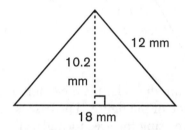

Note that the height must be perpendicular to the base, so the height is 10.2 mm and the base is 18 mm.

$A = \frac{1}{2}bh$ Substitute in the given information.

$A = \frac{1}{2} \times 18 \times 10.2$ Multiply the terms on the right together.

$A = 91.8$ Include the square units.

$A = 91.8$ mm²

CALCULATOR TIPS

When calculating areas, the calculator may help. If you have a fraction key on your calculator, use it when calculating the area of a triangle. The key strokes are shown below to calculate the area of the triangle in the above example: $A = \frac{1}{2} \times 18 \times 10.2$

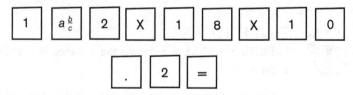

Example:
Find the area of the parallelogram.

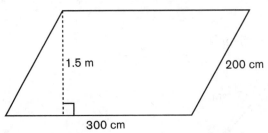

Because the figure is a parallelogram, the height is the length that is perpendicular to the base, not a side of the figure. The base is 300 cm, and the height is 1.5 m. Before using the area formula, all units need to be consistent. Change 300 cm into meters before proceeding. There are 100 centimeters in a meter; therefore, there are 300 divided by 100 meters, which is 3 meters in the base.

$A = bh$	Use the area formula and substitute in the given lengths.
$A = 1.5 \times 3$	Multiply the base times the height.
$A = 4.5$	Include the square units.
$A = 4.5 \text{ m}^2$	

SHORTCUT
The area of a square is $A = s^2$, where s is a side of the square.

PERIMETER AND AREA OF A CIRCLE

RULE BOOK
CIRCUMFERENCE of a circle is the distance AROUND the circle (the perimeter).

$C = \pi d$, where π is a constant, and d is the length of the diameter.
OR
$C = 2\pi r$, where r is the length of the radius.

AREA of a circle is the number of square units it takes to cover the circle.

$A = \pi r^2$, where π is a constant and r is the radius.

PI, π, is a special ratio that is a constant value of approximately 3.14. Pi compares circumference to diameter in the following ratio: $\pi = \frac{C}{d}$. It is the same value for every circle. Often, in math tests, answers will be given in terms of π, such as 136π square units. If answers are not given in terms of π, use the π key on your calculator unless otherwise instructed. Sometimes, a problem will direct you to use either π = 3.14, or $\pi = \frac{22}{7}$, which are approximations for pi.

CALCULATOR TIPS

Many calculators have a special key for the constant pi, which is more accurate than using either π = 3.14, or $\pi = \frac{22}{7}$. Other calculators have the constant as a second function, or shift function. In this case, look for the π symbol printed above one of the keys. Usually, the key looks like one of the following:

Using the formulas above, you can calculate the circumference and area of circles. Take care and check if the problem gives the radius or diameter. If the problem asks for the area of the circle, for example, and gives the length of the diameter, you must first calculate the length of the radius. The radius can be found by dividing the diameter by two. Just like for all area calculations, the units will be square units. The units for circumference will be linear (single) units.

Example:
Find the area of the circle.

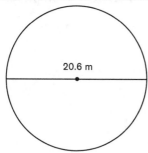

The problem gives the diameter, which is 20.6 meters. The first step is to calculate the radius.

$r = \frac{1}{2} d$ Substitute in 20.6 for *d*.

$r = 10.3$ m Now, use the formula for the area of a circle.

$A = \pi r^2$ Substitute in for the radius.

$A = \pi \times 10.3 \times 10.3$ Multiply 10.3 times 10.3, and include the square units.

$A = 106.09\pi$ m²

Frequently, answers are left in terms of π, as in this example. Take care on multiple-choice tests. Sometimes the answer choices will look similar; one will be in terms of π, and another will omit the π constant.

Example 2:
Given that the circumference of a circle is $C \approx 106.76$ feet, find the radius. Use π = 3.14.

$C = \pi d$	Substitute in for C and π.
$106.76 = 3.14 \times d$	Divide 106.76 by 3.14 to find diameter.
$34 = d$	To find the radius, use the radius formula.
$r = \frac{1}{2}d$	Substitute in the diameter.
$r = \frac{1}{2} \times 34$	Multiply one-half times 34, and include units.
$r = 17$ ft	

AREA OF IRREGULAR SHAPED FIGURES

Some problems ask for the area of an irregular shaped polygon. The key to solving these types of problems is to break the figure up into polygons that you recognize, such as a triangle, rectangle, or circle. Often, pieces of circles are part of the irregular figure.

Example:
Find the area to the nearest hundredth.

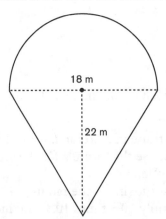

Notice that this figure is a triangle with one-half of a circle on the top. The area is $A = A_{\text{triangle}} + \frac{1}{2}A_{\text{circle}}$. The base of the triangle is equal to the diameter of the circle, so the radius of the circle is one half the base of the triangle. The radius is 9. Substitute in the known formulas: $(\frac{1}{2} \times 18 \times 22) + \frac{1}{2} \times \pi \times 9^2$. Multiply first and then add to get $198 + 127.28 = 325.23$ m².

127.17   325.17

Example:
Find the area of the figure. Use 3.14 for π.

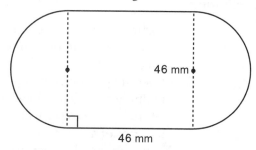

46 mm

46 mm

This figure is actually a square; with two half circles on either end. The two half circles are congruent, and together they form a whole circle. The radius of the circle is one-half of the side of the square, and one-half of 46 is 23 mm. $A = A_{\text{square}} + A_{\text{circle}}$. Substitute in the formulas: $A = s^2 + \pi r^2$. $A = (46 \times 46) + (3.14 \times 23 \times 23)$. Multiply first, and then add to get the area. $A = 2{,}116 + 1{,}661.06 = 3{,}777.06$ mm².

Example:
Find the perimeter of the above shape.

In this case, the perimeter will be two sides of the square, plus the circumference of one whole circle; $P = 46 + 46 + \pi d$; $P = 92 + (3.14 \times 46)$; $P = 92 + 144.44 = 236.44$ mm.

AREA OF SHADED REGIONS

Often on math tests, you are asked to find the area of a shaded region, such as these:

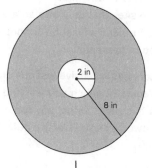

I

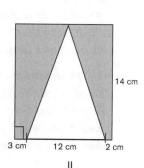

II

To solve this type of problem, you identify the figures in the diagram. There is an outer figure and an inner figure. The area of the shaded region will be Area$_{\text{outer figure}}$ minus the Area$_{\text{inner figure}}$.

In Figure I, the outer figure is a circle with a radius of 8 inches, and the inner figure is a circle with a radius of 2 inches. To find the area of the shaded region, perform the following:

$A_{\text{shaded}} = A_{\text{outer}} - A_{\text{inner}}$ — Substitute in the correct formulas.

$A_{\text{shaded}} = \pi r^2 - \pi r^2$ — Now, substitute in the radius lengths.

$A_{\text{shaded}} = (\pi \times 8 \times 8) - (\pi \times 2 \times 2)$ — Order of operations directs multiplication to be done next, left to right. Answer will be left in terms of π.

$A_{\text{shaded}} = 64\pi - 4\pi$ — Now, combine the π terms, and include square units.

$A_{\text{shaded}} = 60\pi \text{ in}^2$

In Figure II, the outer figure is a rectangle and the inner figure is a triangle. The height of both the rectangle and the triangle is 14 cm. The base of the rectangle is 17 cm and the base of the triangle is 12 cm.

$A_{\text{shaded}} = A_{\text{outer}} - A_{\text{inner}}$ — Substitute in the correct formulas.

$A_{\text{shaded}} = bh - \frac{1}{2}bh$ — Substitute in the given lengths.

$A_{\text{shaded}} = (17 \times 14) - (\frac{1}{2} \times 12 \times 14)$ — Multiplication is done next, working left to right.

$A_{\text{shaded}} = 238 - 84$ — Now evaluate the subtraction, and include the units.

$A_{\text{shaded}} = 154 \text{ cm}^2$

EXTRA HELP

If you feel you need extended help in working with area and perimeter, Geometry Success in 20 Minutes a Day, published by LearningExpress, has several lessons devoted to this topic.

There are several useful web sites that deal with area and perimeter. Visit these sites if you need further clarification on these concepts. Each one has a unique method of presentation.

1. The website *www.math.com* has extensive lessons on geometry. Once at the site, click on *Geometry*, which you will find on the left under *Select Subject*. From this page, select any topic of interest. Each topic has a lesson, followed by an interactive quiz. Answers to all quizzes are provided.

2. The website *www.aaamath.com* is another good resource for practice with geometry. Once on the home page, click on *Geometry*. You will find this on the right under *Math Topics*. The topics are well organized, and there is a brief description of the topic followed by an interactive quiz. Answers are provided.

3. A third site to visit is *http://rock.uwc.edu/galexand/baw/toc.htm*. You will see the heading *Table of Contents* on the left. Click on *Lesson* #10, Geometric Figures. Again, there is a lesson followed by a quiz with answers provided.

TIPS AND STRATEGIES

- Perimeter is a linear measurement that measures the length around a plane figure.
- Area is a square measurement that measures how many square units it takes to cover a figure.
- Be careful of unit consistency when solving problems of area and perimeter.
- Circles can be measured for circumference and area.
- Know the formulas for the area of a circle and common polygons.
- Calculate the area of irregular-shaped figures by breaking them up into smaller figures that are recognizable.
- Calculate the area of a shaded region by subtraction:
  $A_{shaded} = A_{outer} - A_{inner}$.

PRACTICE QUIZ

1. Find the area.

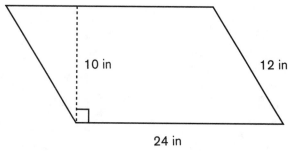

24 in

    a. 160 in²
    b. 46 in²
    c. 120 in²
    d. 288 in²
    e. 240 in²

2. Find the perimeter.

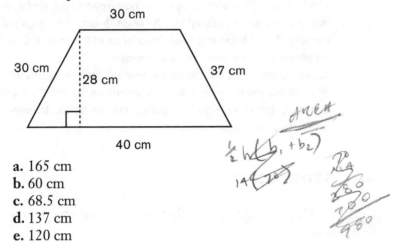

30 cm

30 cm   28 cm   37 cm

40 cm

a. 165 cm
b. 60 cm
c. 68.5 cm
d. 137 cm
e. 120 cm

3. Find the perimeter.

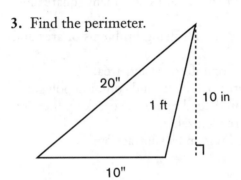

20"   1 ft   10 in

10"

a. 52 in
b. 42 in
c. 31 ft
d. 3.1 ft
e. 50 in

4. Find the area.

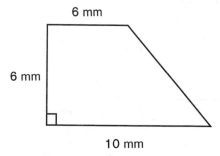

6 mm

6 mm

10 mm

    a. 36 mm²
    b. 48 mm²
    c. 22 mm²
    d. 64 mm²
    e. 60 mm²

5. Find the area.

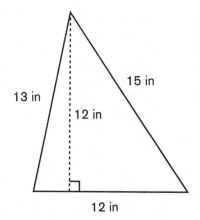

15 in

13 in

12 in

12 in

    a. 24 in²
    b. 40 in²
    c. 72 in²
    d. 144 in²
    e. 52 in²

6. Find the perimeter.

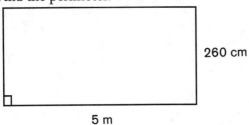

260 cm

5 m

a. 1,520 cm
b. 1,520 m
c. 265 cm
d. 265 m
e. 1,300 cm

7. Find the area.

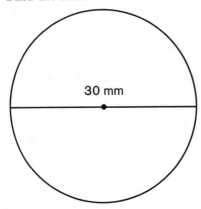

30 mm

a. 900π mm^2
b. 60 mm^2
c. 225π mm^2
d. 900 mm^2
e. 225 mm^2

8. Find the perimeter of the equilateral triangle.

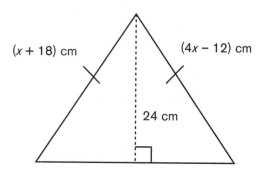

$(x + 18)$ cm $(4x - 12)$ cm

24 cm

  a. 28 cm
  b. 56 cm
  c. 84 cm
  d. 80 cm
  e. 336 cm

9. Given the circumference of a circle is 24π inches, find the radius of the circle.

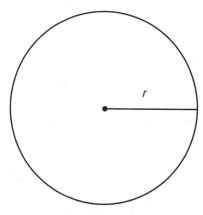

$C = 24\pi$ inches

  a. 6 in
  b. 6π in
  c. 12 in
  d. 12π in
  e. 24 in

10. Find the circumference.

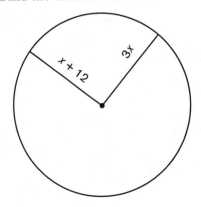

    a. 6 cm
    b. 6π cm
    c. 18π cm
    d. 9π cm
    e. 36π cm

11. Find the area of the square.

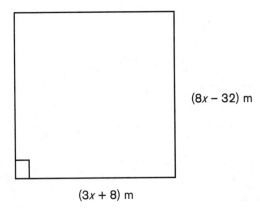

    a. 8 m²
    b. 32 m²
    c. 64 m²
    d. 128 m²
    e. 1,024 m²

12. Given the area of the rectangle = 90 cm², find the length of $\overline{AB}$.

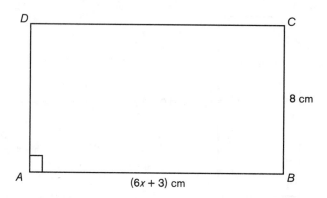

a. 15 cm
b. 11.25 cm
c. 18 cm
d. 2.5 cm
e. 8 cm

13. Find the perimeter.

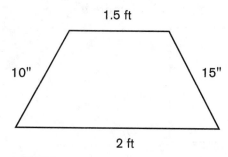

a. 28.5 in
b. 67 in
c. 64 in
d. 55 in
e. 46 in

14. Find the area of the rectangle.

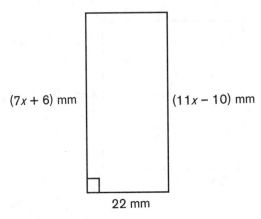

$(7x + 6)$ mm $(11x - 10)$ mm

22 mm

a. 748 mm²
b. 88 mm²
c. 112 mm²
d. 56 mm²
e. 52 mm²

15. Find the area of the figure to the nearest hundredth.

12 cm

15 cm

a. 180 cm²
b. 236.55 cm²
c. 293.10 cm²
d. 632.29 cm²
e. 406.19 cm²

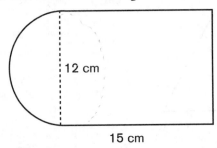

16. Find the area of the figure to the nearest hundredth.

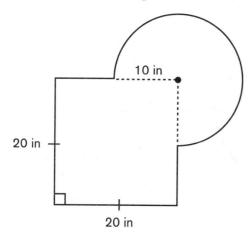

    a. 400 in²
    b. 714.16 in²
    c. 462.83 in²
    d. 635.62 in²
    e. 478.54 in²

17. Find the area of the shaded region. Use 3.14 for π.

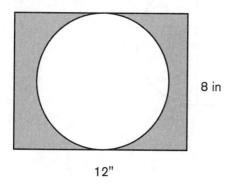

    a. 45.76 in²
    b. 105.06 in²
    c. 70.87 in²
    d. 146.27 in²
    e. 297.06 in²

18. Find the area of the shaded region.

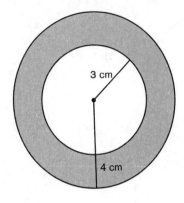

a. 58π cm²
b. π cm²
c. 7π cm²
d. 25π cm²
e. 40π cm²

19. Find the area of the figure to the nearest hundredth.

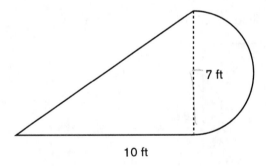

a. 54.24 ft²
b. 111.97 ft²
c. 73.48 ft²
d. 108.48 ft²
e. 146.97 ft²

20. Find the area of the figure.

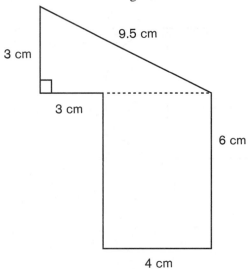

- **a.** 31.5 cm²
- **b.** 28.5 cm²
- **c.** 34.5 cm²
- **d.** 33 cm²
- **e.** 39 cm²

21. Find the area of the shaded region.

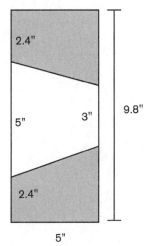

- **a.** 49 in²
- **b.** 69 in²
- **c.** 29 in²
- **d.** 80 in²
- **e.** 9 in²

22. Find the perimeter. Use π = 3.14.

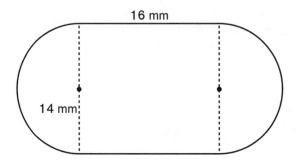

16 mm

14 mm

a. 53.98 mm
b. 107.98 mm
c. 185.94 mm
d. 59.96 mm
e. 75.96 mm

23. The radius of the circle in the figure below is 5 cm. Find the area of the shaded region.

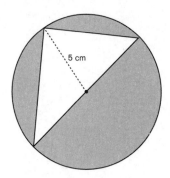

5 cm

a. 157 cm²
b. 28.5 cm²
c. 47.1 cm²
d. 53.5 cm²
e. 103.5 cm²

24. Find the area of the figure.

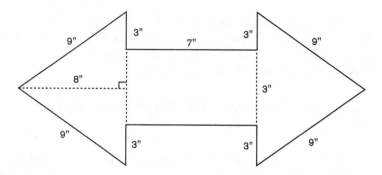

    a. 93 in²
    b. 34.5 in²
    c. 28 in²
    d. 75 in²
    e. 57 in²

25. Find the area of the shaded region.

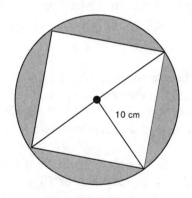

    a. (200π − 100) cm²
    b. (100π − 100) cm²
    c. (40π − 200) cm²
    d. (100π − 200) cm²
    e. (200π − 200) cm²

ANSWERS

1. e. The area of a parallelogram is $A = bh$, where b is the base and h is the height. The side measurement of 12 inches is not needed to solve this problem. Substitute in the values for the base and height: $A = 24 \times 10 = 240$ square inches.

2. d. The perimeter is the sum of the measure of the sides of a polygon; $30 + 30 + 40 + 37 = 137$ cm.

3. b. The perimeter is the sum of the measure of the sides of a polygon. One of the sides is given as feet, and the other two sides are given as inches. Convert 1 foot to 12 inches; $20 + 10 + 12 = 42$ inches.

4. b. The area of a trapezoid is $A = \frac{1}{2}h(b_1 + b_2)$, where h is the height and b_1 and b_2 are the parallel bases. For this trapezoid, the height is also the side perpendicular to the parallel bases. Substitute in to get: $A = \frac{1}{2} \times 6 (6 + 10) = 3 \times 16 = 48$ mm².

5. c. The area of a triangle is $A = \frac{1}{2}bh$, where b is the base and h is the height. Substitute in to get: $A = \frac{1}{2} \times 12 \times 12 = 6 \times 12 = 72$ square inches.

6. a. The units must be consistent before solving. Change 5 meters to 500 cm. This is a rectangle, so use the shortcut formula: $P = 2l + 2w$. Substitute in to get: $(2 \times 260) + (2 \times 500) = 520 + 1,000 = 1,520$ cm. Note that choice **b** has units of meters, which is incorrect.

7. c. The area of a circle is $A = \pi r^2$. This problem gives the diameter. The radius is one-half of the diameter. One-half of 30 is 15 mm. Substitute in to get: $A = \pi \times 15^2 = 225\pi$ square millimeters. Note that choice **e** is just 225 mm², which is incorrect.

8. c. An equilateral triangle has all three sides congruent, of equal measure. Use algebra to find the value of the variable x. Find the length of one of the sides and multiply by three:

$x + 18 = 4x - 12$	Set up an equation.
$x + 18 - x = 4x - 12 - x$	Subtract x from both sides.
$18 = 3x - 12$	Combine like terms.
$18 + 12 = 3x - 12 + 12$	Add 12 to both sides.
$30 = 3x$	Combine like terms.
$\frac{30}{3} = \frac{3x}{3}$	Divide both sides by 3.
$10 = x$	

Use the value of 10 to find the length of side $x + 18$: $10 + 18 = 28$ cm. The perimeter is $3 \times 28 = 84$ cm.

9. c. The radius is one-half of the diameter of a circle. Since the formula for circumference is $C = \pi d$, the diameter is 24 inches. One-half of 24 equals 12, so the radius is 12 inches.

10. e. The circumference is $C = \pi d$, where d is the diameter of the circle. Two radii are shown, and all radii are congruent in a circle. Use algebra to solve for the variable x, to find the diameter, then solve for the circumference:

$3x = x + 12$	Set up the equation.
$3x - x = x + 12 - x$	Subtract x from both sides.
$2x = 12$	Combine like terms.
$\frac{2x}{2} = \frac{12}{2}$	Divide both sides by 2.
$x = 6$	

Use the value of 6 to find the diameter, which is two times a radius: $2(3x) = 2 \times (3 \times 6) = 2 \times 18 = 36$. The circumference is 36π cm.

11. e. A square has all four sides of equal measure. Use algebra to find the value of x, and then use the shortcut for the area of a square, $A = s^2$.

$8x - 32 = 3x + 8$	Set up the equation.
$8x - 32 - 3x = 3x + 8 - 3x$	Subtract $3x$ from both sides.
$5x - 32 = 8$	Combine like terms.
$5x - 32 + 32 = 8 + 32$	Add 32 to both sides.
$5x = 40$	Combine like terms.
$\frac{5x}{5} = \frac{40}{5}$	Divide both sides by 5.
$x = 8$	

Use the value of 8 to find the length of one of the sides of the square: $3x + 8 = (3 \times 8) + 8 = 24 + 8 = 32$ meters. The area is $32^2 = 1,024$ m^2.

12. b. The area is given as 90 cm², and the formula for the area is $A = bh$. Use algebra to solve for the variable x, and then use this value to find the length of $\overline{AB}$.

$90 = 8(6x + 3)$	Set up the equation.
$90 = (8 \times 6x) + 24$	Use the distributive property.
$90 - 24 = 48x + 24 - 24$	Subtract 24 from both sides.
$66 = 48x$	Combine like terms.
$\frac{66}{48} = \frac{48x}{48}$	Divide both sides by 48.
$1.375 = x$	

Use this value to find the length of $\overline{AB}$: $6x + 3 = (6 \times 1.375) + 3 = 8.25 + 3 = 11.25$ cm.

13. b. To find the perimeter, add up the measure of the sides. First make all units consistent. All of the answer choices are inches, so convert the feet measurements to inches by multiplying by 12: 2 feet = 24 inches, and 1.5 feet = 18 inches. Perimeter is 24 + 18 + 10 + 15 = 67 inches.

14. a. The area of a rectangle is $A = bh$. The opposite sides of a rectangle are equal measure, so use algebra to find the variable of x:

$7x + 6 = 11x - 10$	Set up the equation.
$7x + 6 - 7x = 11x - 10 - 7x$	Subtract $7x$ from both sides.
$6 = 4x - 10$	Combine like terms.
$6 + 10 = 4x - 10 + 10$	Add 10 to both sides.
$16 = 4x$	Combine like terms.
$\frac{16}{4} = \frac{4x}{4}$	Divide both sides by 4.
$4 = x$	

Use this value to find the length of the longer sides: $7x + 6 = (7 \times 4) + 6 = 28 + 6 = 34$ mm. Now, calculate the area = $34 \times 22 = 748$ mm².

15. b. This irregular figure is a rectangle and one-half of a circle. The side of the rectangle, which is also the diameter of the circle, is 12 cm. The radius is one half the measure of the diameter so the radius is 6 cm; $A_{\text{irregular figure}} = A_{\text{rectangle}} + A_{\frac{1}{2}\,\text{circle}}$. The answer calls for the value to the nearest hundredth and does not specify what value to use for π. Use the π key on the calculator. Substitute in the formulas and then the values: $A = bh + \frac{1}{2}\pi r^2 = (15 \times 12) + \frac{1}{2}\pi 6^2 = 180 + (18 \times \pi) = 180 + 56.55 = 236.55$ square centimeters.

16. d. This irregular figure is a square with sides of 20 inches, and three fourths of a circle with radius of 10 inches. Add up the two pieces to get the total area; $A = s^2 + \frac{3}{4}\pi r^2$. Substitute in the values to get: $20^2 + \frac{3}{4}\pi 10^2$. Use the π key to solve: $400 + \frac{3}{4} \times 100 \times \pi = 400 + 235.62 = 635.62$ in².

17. a. The shaded region is $A_{shaded} = A_{outer} - A_{inner}$. The outer shape is a rectangle and the inner shape is a circle. The radius of the circle is one-half the height of the rectangle, or one half of 8, which is 4 inches; $A = bh - \pi r^2$. Substitute in to get $A = (12 \times 8) - (3.14 \times 4^2)$ or $96 - 50.24 = 45.76$ square inches.

18. e. The area of the shaded region is $A_{shaded} = A_{outer} - A_{inner}$. Two segment lengths are given. One is the radius of the inner circle, the other measure, 4 cm, is the distance from the inner circle to the outer circle. The outer figure is actually a circle with a radius of 7 and the inner figure is a circle with a radius of 3. All of the answer choices are left in terms of π; $A = \pi r^2 - \pi r^2$. Substitute in to get: $A = 7^2\pi - 3^2\pi = 49\pi - 9\pi = 40\pi$ cm².

19. a. This irregular figure is a right triangle, with base of 10 and height of 7, and a half circle with radius of 3.5; $A = \frac{1}{2}bh + \frac{1}{2}\pi r^2$; $A = (\frac{1}{210} \times 7) + (\frac{1}{2}\pi \times 3.5^2)$. Use your calculator, and the π key, to get: $A = 35 + 19.24 = 54.24$ ft².

20. c. This figure is a right triangle and a rectangle. The rectangle has base of 4 and height of 6. The triangle has base of 3 plus the base of the rectangle $(3 + 4 = 7)$ and height of 3. Substitute these values into the formulas: $A = bh + \frac{1}{2}bh = (4 \times 6) + \frac{1}{2}(7 \times 3)$. Simplify to get: $24 + 10.5 = 34.5$ cm².

21. c. The area of the shaded region is $A_{shaded} = A_{outer} - A_{inner}$. The outer figure is a rectangle and the inner figure is a trapezoid. The rectangle has a base of 5 and a height of 9.8. The trapezoid has a height of 5 and the bases are 5 and 3. Substitute in to get: $A = (9.8 \times 5) - \frac{1}{25} \times (5 + 3)$. Simplify to get: $49 - 20 = 29$ in².

22. e. The perimeter of this irregular shaped figure is the sum of the two sides of the rectangle (length 16 mm), and the circumference of two one-half circles with a diameter of 14 mm. Since it is two one-half identical circles, it will just be the circumference of one circle with diameter of 14 mm; $P = 16 + 16 + \pi d = 32 + 3.14 \times 14 = 32 + 43.96 = 75.96$ mm.

23. d. The area of the shaded region is $A_{\text{shaded}} = A_{\text{outer}} - A_{\text{inner}}$. The outer figure is a circle with a radius of 5 cm. The inner figure is a triangle. The base of the triangle is a diameter of the circle; the base is a chord that passes through the center of the circle. The base is 10 cm, which is twice the radius, and the height is 5 cm, the radius of the circle; $A = \pi r^2 - \frac{1}{2}bh = (\pi \times 5^2) - (\frac{1}{2} \times 10 \times 5)$. Simplify to get $78.5 - 25 = 53.5$ cm^2.

24. a. The area of this irregular figure is the sum of the areas of two identical triangles, plus the area of the middle section, a rectangle. The triangles have a base of 9 in and a height of 8 in. The rectangle has a base of 7 in and a height of 3 in; $A = (2 \times (\frac{1}{2} \times 9 \times 8)) + (7 \times 3)$. This simplifies to $72 + 21 = 93$ square inches.

25. d. The area of the shaded region is $A_{\text{shaded}} = A_{\text{outer}} - A_{\text{inner}}$. The outer figure is a circle, and the inner figure is a rectangle. The base and height of the rectangle are not given, so treat this rectangle as two triangles, where the base is the diagonal of the rectangle, which happens to be the diameter of the circle. The radius is given as 10 cm, so the base of the triangles is 20 cm. The height of the triangles is the radius, 10 cm. Substitute these values in to get: $A = \pi 10^2 - 2(\frac{1}{2}bh)$. Simplify to get: $100\pi - (2 \times \frac{1}{2} \times 200)$, or $(100\pi - 200)$ cm^2.

7

Surface Area and Volume

Chapter 6 reviewed the concepts of measurement in two dimensions. This chapter concentrates on the three-dimensional measures of surface area and volume. Take this ten-question benchmark quiz to assess your current knowledge of these concepts.

BENCHMARK QUIZ

1. How many edges does this rectangular prism have?

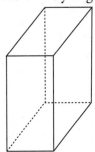

   a. 9
   b. 12
   c. 6
   d. 8
   e. 11

2. Find the surface area of the triangular prism.

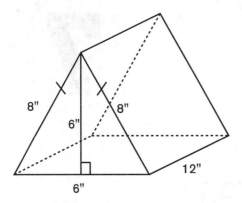

   a. 360 in²
   b. 468 in²
   c. 342 in²
   d. 300 in²
   e. 324 in²

3. Find the volume of the trapezoidal prism.

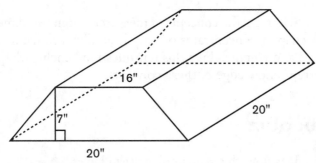

   a. 2,520 in³
   b. 2,800 in³
   c. 63 in³
   d. 6,400 in³
   e. 5,040 in³

4. Find the surface area.

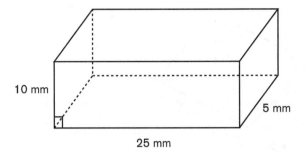

10 mm

5 mm

25 mm

   a. 1,250 mm²
   b. 1,100 mm²
   c. 850 mm²
   d. 1,000 mm²
   e. 2,500 mm²

5. Find the surface area.

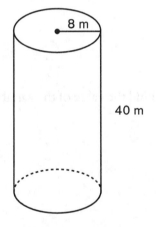

8 m

40 m

   a. 704π m²
   b. 704 m²
   c. 768 m²
   d. 656 m²
   e. 768π m²

6. The base of the triangular prism is an isosceles right triangle. Find the volume.

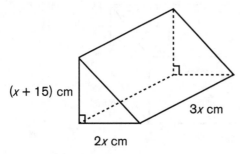

$(x + 15)$ cm

$3x$ cm

$2x$ cm

    a. 20,250 cm³
    b. 40,500 cm³
    c. 342 cm³
    d. 450 cm³
    e. 144 cm³

7. Find the surface area of a sphere whose diameter is 14 meters.
    a. 14π m²
    b. 49π m²
    c. 196π m²
    d. 784π m²
    e. 49 m²

8. The volume of the cylinder is 384π in³. Find the value of the variable x.

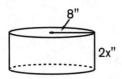

8"

$2x$"

    a. 19.2 in
    b. 3 in
    c. 6 in
    d. 21.5 in
    e. 43 in

9. Find the volume of the sphere.

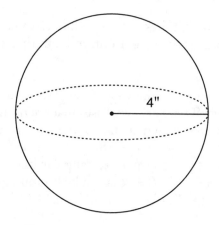

4"

a. $\frac{64\pi}{3}$ in^3

b. 64π in^3

c. $\frac{128\pi}{3}$ in^3

d. 256π in^3

e. $\frac{256\pi}{3}$ in^3

10. Find the volume of the cone.

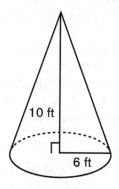

10 ft

6 ft

a. 360π ft^3

b. 30π ft^3

c. 60π ft^3

d. 120π ft^3

e. 90π ft^3

BENCHMARK QUIZ SOLUTIONS

1. b. The edges are the segments that make up the solid. There are 12 segments: four on the top, four on the bottom, and four that make up the sides.

2. d. Surface area is found by calculating the area of each face, and then adding these areas together. There are two bases that are triangles of equal measure, and three rectangular faces. The area of each triangle is $A = \frac{1}{2}bh$, or $A = \frac{1}{2} \times 6 \times 6 = 18$ in². Since this is an isosceles triangle, there are two congruent rectangular faces. The area of each is $A = bh$, or $A = 12 \times 8 = 96$ in². The bottom face is $A = bh$, or $A = 12 \times 6 = 72$ in². Surface area is $18 + 18 + 96 + 96 + 72 = 300$ in².

3. a. Volume of a prism is found by $V = Bh$, where B is the area of the base. In this prism, the base is a trapezoid; $V = \frac{1}{2}h_1(b_1 + b_2) \times h_2$. Substitute in the values for the heights and bases: $V = \frac{1}{2} \times 7(20 + 16) \times 20$, $V = \frac{1}{2} \times 7(36) \times 20 = 2{,}520$ in³.

4. c. The surface area of a rectangular prism can be found by using the formula $SA = 2(l \times w) + 2(l \times h) + 2(w \times h)$. Substitute in the values for length, width, and height to get: $2(25 \times 5) + 2(25 \times 10) + 2(5 \times 10) = 2(125) + 2(250) + 2(50) = 250 + 500 + 100 = 850$ mm².

5. e. Use the formula for the surface area of a cylinder: $SA = 2(\pi r^2) + 2\pi rh$. Substitute in 8 for r, the radius, and 40 for h, the height; $SA = 2(\pi 8^2) + 2\pi \times 8 \times 40$. Evaluate the exponent: $2(\pi 64) + 2\pi \times 8 \times 40$. Perform multiplication: $128\pi + 640\pi$. Combine like terms to get the surface area of 768π m². Be careful; choice **c** is 768 m², which is incorrect.

6. a. The base is a right isosceles triangle. The base and height have equal measure. Use this fact and algebra to find the value of the variable x. Evaluate the length of the dimensions using this value, and then calculate the volume.

$2x = x + 15$	Set up an equation.
$2x - x = x + 15 - x$	Subtract x from both sides.
$x = 15$	

The base of the triangle, b, is 2 times 15, or 30 cm. The height of the triangle, h_1, is also 30 cm. The height of the prism, h_2, is 3 times 15, or 45 cm. Use the formula $V = \frac{1}{2}bh_1 \times h_2$. Substitute in the measures: $V = \frac{1}{2} \times 30 \times 30 \times 45$. Multiply to get 20,250 cm³.

7. c. The surface area of a sphere is $SA = 4\pi r^2$. The diameter is given in the problem. The radius is one half the length of the diameter, or 7 meters; $SA = 4 \times \pi \times 7^2 = 4(49)\pi$, or 196π square meters.

8. b. Use the formula for the volume of a cylinder and an algebraic equation to find the value of the variable x. The volume of a cylinder is $V = \pi r^2 h$.

$\pi(8^2)(2x) = 384\pi$	Set up an equation.
$\pi(128x) = 384\pi$	Multiply.
$\dfrac{128\pi x}{128\pi} = \dfrac{384\pi}{128\pi}$	Divide both sides by 128π.
$x = 3$	

9. e. Use the formula for the volume of a sphere: $V = \frac{4}{3}\pi r^3$. This sphere has a radius of 4 inches; $V = \frac{4}{3}\pi 4^3 = \frac{4}{3} \times \pi \times 64$. Multiply to get $\frac{256\pi}{3}$ in^3.

10. d. The volume of a cone is $V = \frac{1}{3}\pi r^2 h$, where r is the radius of the base, and h is the height of the cone. Substitute in the values for radius and height; $V = \frac{1}{3}\pi 6^2 \times 10$. Evaluate the exponent: $V = \frac{1}{3}\pi \times 36 \times 10$ and multiply to get 120π ft^3.

BENCHMARK QUIZ RESULTS

If you answered 8–10 questions correctly, you have remembered a great deal about surface area and volume. If the questions you answered incorrectly are centered on one specific area in this chapter, take your time on that section in the chapter. Read over the remainder of the text to pick up any shortcuts or techniques for solving problems. Then, proceed to the Practice Quiz to try to improve your score.

If you answered 4–7 questions correctly, there are several concepts you need to review. Carefully read through the lesson in this chapter for review and skill building. Work carefully through the examples and pay attention to the sidebars that refer you to definitions, hints, and shortcuts. Get additional practice on geometry by visiting the suggested websites and taking the quiz at the end of the chapter.

If you answered 1–3 questions correctly, carefully read this chapter and concentrate on the sidebars and visual aids that will help with comprehension. Work through each example in the text on paper to be sure you can solve it. Go to the suggested websites in the Extra Help sidebar in this chapter, which will help with understanding and will provide extended practice. Refer to *Geometry Success in 20 Minutes a Day*, published by LearningExpress. This book has lessons 14, 15, and 16 that cover these concepts.

JUST IN TIME LESSON— SURFACE AREA AND VOLUME

There are many applications of geometry in mathematical problems. This chapter will review three-dimensional solid figures. Surface area is a *covering* or *wrapping* concept and volume is a *filling* concept.

The topics in this chapter are:

- Three-dimensional Figures: Faces, Vertices, and Edges
- Surface Area of Prisms
- Surface Area of Cylinders and Spheres
- Volume of Prisms
- Volume of Cylinders and Spheres
- Volume of Pyramids and Cones

THREE-DIMENSIONAL FIGURES: FACES, VERTICES, AND EDGES

Solid figures are three-dimensional entities. Polyhedrons are solids whose surfaces are made up of polygons. The parts of a polyhedron are defined as *faces*, *vertices*, and *edges*.

 GLOSSARY

FACE of a polyhedron is one of the plane surfaces on the solid
EDGE of a polyhedron is one of the segments on the solid
VERTEX of a polyhedron is an intersecting point of any two edges of the solid

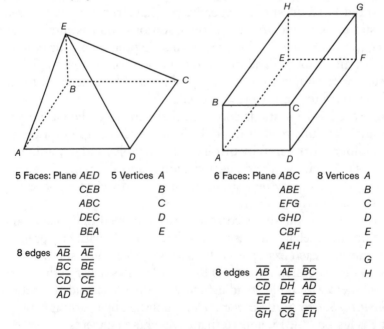

5 Faces: Plane *AED*	5 Vertices *A*
CEB	B
ABC	C
DEC	D
BEA	E

8 edges	$\overline{AB}$	$\overline{AE}$
	$\overline{BC}$	$\overline{BE}$
	$\overline{CD}$	$\overline{CE}$
	$\overline{AD}$	$\overline{DE}$

6 Faces: Plane *ABC*	8 Vertices *A*
ABE	B
EFG	C
GHD	D
CBF	E
AEH	F
	G
	H

8 edges	$\overline{AB}$	$\overline{AE}$	$\overline{BC}$
	$\overline{CD}$	$\overline{DH}$	$\overline{AD}$
	$\overline{EF}$	$\overline{BF}$	$\overline{FG}$
	$\overline{GH}$	$\overline{CG}$	$\overline{EH}$

Some common solid figures are defined below.

GLOSSARY

PRISM a three dimensional solid that has two congruent faces called *bases*. The other faces are rectangles.

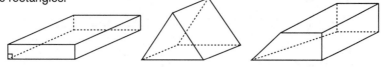

RECTANGULAR PRISM a prism that has bases that are rectangles
CUBE a rectangular prism with six congruent faces that are squares

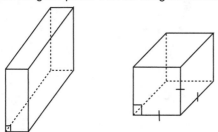

TRIANGULAR PRISM a prism that has bases that are triangles
TRAPEZOIDAL PRISM a prism that has bases that are trapezoids

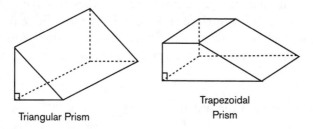

Trapezoidal
Prism

Triangular Prism

CYLINDER is a solid in which the bases are circles and the other surface is a rectangle wrapped around the circles
SPHERE a solid in which all of the points on the sphere are an equal distance from a center point. This distance is the radius of the sphere.

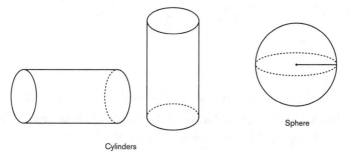

Sphere

Cylinders

RECTANGULAR PYRAMID a solid that has a rectangle for a base, and four triangular faces that meet at a vertex opposite to the base

CONE a solid that has a circle for a base, with the other face wrapping around this base and meeting opposite the base at a vertex point

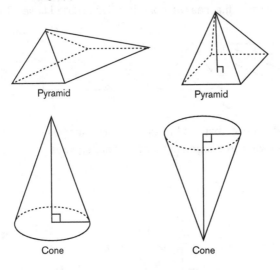

Pyramid   Pyramid

Cone   Cone

SURFACE AREA OF PRISMS

Surface area is the number of square units that it takes to cover a three dimensional solid. To calculate surface area, first determine the number of faces on the prism. Calculate the area of each face and then add them together. The formulas for the area of the faces were covered in Chapter 6.

Example:
Find the surface area of the prism.

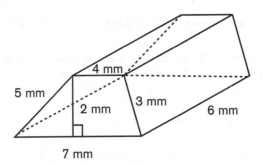

5 mm

4 mm

2 mm

3 mm

6 mm

7 mm

This is a trapezoidal prism; the parallel faces are trapezoids. There are six faces: two congruent trapezoids, and four rectangles. The area of each trapezoid is $A = \frac{1}{2}h(b_1 + b_2)$, or $A = \frac{1}{2} \times 2 \times (7 + 4) = 11$.

The area of the bottom rectangle is $A = bh$, or $A = 7 \times 6 = 42$. The area of the top face is $A = 4 \times 6 = 24$. The area of the left face is $A = 5 \times 6 = 30$, and the area of the right face is $A = 3 \times 6 = 18$. Add these six areas together to get the surface area: $SA = 11 + 11 + 42 + 24 + 30 + 18 = 136$ mm².

Example:
The surface area of the triangular prism shown is 242.4 in². The triangular base is an equilateral triangle. What is the length of $\overline{BC}$?

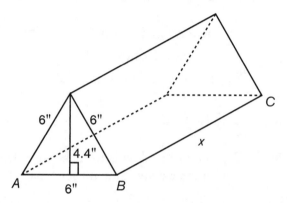

Use the variable x to represent the length of $\overline{BC}$. The area of the two triangular bases is $A = \frac{1}{2}bh$, or $A = \frac{1}{2} \times 6 \times 4.4 = 13.2$. There are three other rectangular faces. Each of these is congruent because the triangular base is equilateral. The area of each rectangle is $A = bh$, or $A = 6x$. Use algebra to find the value of the variable x, the length of $\overline{BC}$:

$242.4 = 13.2 + 13.2 + 3(6x)$	Set up the equation.
$242.4 = 26.4 + 18x$	Multiply, and then combine like terms.
$242.4 - 26.4 = 26.4 + 18x - 26.4$	Subtract 26.4 from both sides.
$216 = 18x$	Combine like terms.
$\frac{216}{18} = \frac{18x}{18}$	Divide both sides by 18.
$12 = x$	This is the length of $\overline{BC}$.

SHORTCUT

For prisms, there is a pair of parallel congruent bases. Calculate the area of one of these bases and then multiply by 2. For rectangular prisms, there are two other sets of congruent faces. These facts lead to the formulas commonly used for rectangular prisms:

SURFACE AREA OF A RECTANGULAR PRISM:

$SA = 2(l \times w) + 2(l \times h) + 2(w \times h)$

where w is the width, l is the length, and h is the height of the prism

SURFACE AREA OF A CUBE:

$SA = 6s^2$

where s is the length of a side of the cube

Example:
The surface area of a cube is 384 ft². What is the length of an edge of the cube?

$384 = 6s^2$	Use the formula to set up an equation.
$\dfrac{384}{6} = \dfrac{6s^2}{6}$	Divide both sides by 6.
$\sqrt{64} = \sqrt{s^2}$	Take the square root of each side.
$8 = s$	

Each edge of the cube measures 8 feet.

Be alert when working with geometry problems to make sure that the units are consistent. If they are different, a conversion must be made before calculating surface area.

Example:
What is the surface area of the following rectangular prism?

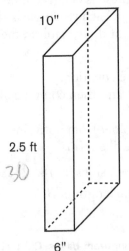

The length is 2.5 feet, the width is 6 inches, and the height is 10 inches. Convert the length to inches (1 foot = 12 inches). The length is 12 × 2.5 = 30 inches. Now use the shortcut formula:
$SA = 2(30 \times 6) + 2(30 \times 10) + 2(6 \times 10)$

$SA = 2(180) + 2(300) + 2(60)$ Evaluate parentheses.
$SA = 360 + 600 + 120$ Perform multiplication.
$SA = 1{,}080 \text{ in}^2$ Perform addition; include units.

When there is a variable present given as one of the dimensions, an algebraic equation is used to either solve for the variable, or to find a measurement. Pay attention to what the problem is asking for; sometimes the value of the variable is needed, other times the actual measurement is requested.

Example:
The triangular base of the following prism is an isosceles triangle. Find the surface area of the prism.

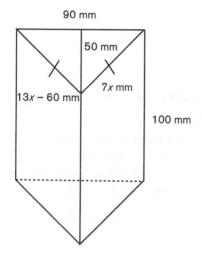

90 mm
50 mm
$7x$ mm
$13x - 60$ mm
100 mm

Since the triangle is isosceles, the sides marked as algebraic expressions have equal measure. Use algebra to solve for the variable x. Evaluate to find the length of these sides. Then, calculate surface area.

$13x - 60 = 7x$ Set up an equation.
$13x - 60 - 7x = 7x - 7x$ Subtract $7x$ from both sides.
$6x - 60 = 0$ Combine like terms.
$6x - 60 + 60 = 0 + 60$ Add 60 to both sides.
$6x = 60$ Combine like terms.
$\frac{6x}{6} = \frac{60}{6}$ Divide both sides by 6.
$x = 1$

The area of the triangle is $A = \frac{1}{2} \times 90 \times 50 = 2{,}250 \text{ mm}^2$. The congruent sides of the triangle are $7x = 7(10) = 70 \text{ mm}^2$. Two of the rectangular faces are $A = 70 \times 100 = 7{,}000 \text{ mm}^2$. The other face is $A =$

$90 \times 100 = 9,000$ mm². Add these areas to find the surface area: $SA = 2(2,250) + 2(7,000) + 9,000 = 4,500 + 14,000 + 9,000 = 27,500$ mm².

Example:
The surface area of the rectangular prism is 652 cm². Find the value of the variable x.

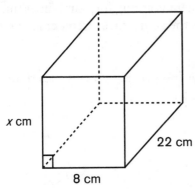

x cm

22 cm

8 cm

Use the shortcut formula and algebra to find the value of x:

$652 = 2(8x) + 2(22x) + 2(8 \times 22)$

$652 = 16x + 44x + 352$ Perform multiplication.

$652 = 60x + 352$ Combine like terms.

$652 - 352 = 60x + 352 - 352$ Subtract 352 from both sides.

$300 = 60x$ Combine like terms.

$\dfrac{300}{60} = \dfrac{60x}{60}$ Divide both sides by 60.

$5 = x$

SURFACE AREA OF CYLINDERS AND SPHERES

A cylinder has two congruent bases that are circles, and one long rectangular piece that wraps to form the side.

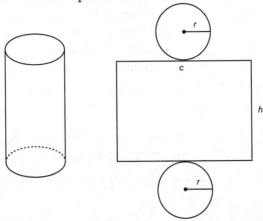

The side marked *h* is the height of the cylinder. The side marked *c* is the circumference of the circular base, calculated by the diameter of the circular base.

 RULE BOOK

The surface area of a cylinder is the two times the area of one of the circular bases, plus the circumference of the circular base, times the height of the cylinder:

SURFACE AREA OF A CYLINDER
$$SA = 2(\pi r^2) + 2\pi rh$$

The surface area of a sphere is based on the radius of the sphere:

SURFACE AREA OF A SPHERE
$$SA = 4\pi r^2$$

Using the preceding formulas, you can calculate the surface area of cylinders and spheres. Take care and check if the problem gives the radius or diameter. If the problem asks for the surface area of a cylinder, for example, and gives the length of the diameter, you must first calculate the length of the radius. The radius can be found by dividing the diameter by 2. Just like for all area calculations, the units will be square units.

Example:
Find the surface area of the following cylinder. Use 3.14 for π.

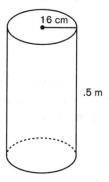

16 cm

.5 m

Note that the units are not consistent. Convert the height from meters to centimeters by multiplying by 100: $0.5 \times 100 = 50$ cm. Use the formula for surface area, substituting 16 for *r* and 50 for *h*:

$SA = 2(\pi 16^2) + 2\pi \times 16 \times 50$.
Evaluate exponents: $2(256\pi) + 2\pi \times 16 \times 50$.
Perform multiplication: $512\pi + 1,600\pi$.
Combine like terms: $SA = 2,112\pi$ cm^2.
Use the value of 3.14 for π to get the surface area: $2,112 \times 3.14 = 6,631.68$ cm^2.

Example:
The surface area of the following cylinder is 384π square units. What is the value of the variable x?

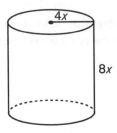

Use the formula and algebra:

$2(\pi16x^2) + 2\pi(4x)(8x) = 384\pi$	Set up the equation.
$32\pi x^2 + 64\pi x^2 = 384\pi$	Perform multiplication.
$96\pi x^2 = 384\pi$	Combine like terms.
$\dfrac{96\pi x^2}{96\pi} = \dfrac{384\pi}{96\pi}$	Divide both sides by 96π.
$\sqrt{x^2} = \sqrt{4}$	Simplify; take the square root of each side.

$x = 2$

Example:
Find the surface area of a sphere with a diameter of 6 inches.

The diameter is given. Calculate the radius, which is one-half the diameter. The radius is 3 inches. Use the formula $A = 4\pi r^2$; $A = 4 \times \pi \times 3^2 = 36\pi$ in².

Frequently, answers are left in terms of π, as in this example. Take care on multiple-choice tests. Sometimes the answer choices will look similar; one will be in terms of π, and another will omit the π constant.

VOLUME OF PRISMS

Volume is a measure of how many cubic units it takes to FILL a solid figure. Volume is measured in cubic units. Volume is a multiplication concept, where three measures are multiplied together. The units can also be thought of as multiplied together: cm $\times$ cm $\times$ cm = cm³, or the words "centimeters cubed." There are formulas to use for the volume of common solid figures.

RULE BOOK

Volume is the area of the base of the solid figure, multiplied by the height of the figure. This can be expressed as $V = Bh$, where V is the volume, B is the area of the base, and h is the height of the prism. The formulas for the area of these base shapes were covered in Chapter 6 of this book.

VOLUME OF A RECTANGULAR PRISM: $V = lwh$

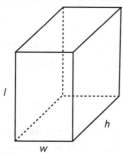

In this case, $B = lw$, (the base is a rectangle), where l is the length, and w is the width.

VOLUME OF A CUBE: $V = s^3$

In this case, $B = s^2$, and the height is also s.

VOLUME OF A TRIANGULAR PRISM: $V = \frac{1}{2}bh_1 \times h_2$

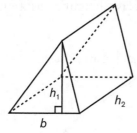

In this case, $B = \frac{1}{2}bh_1$, where b is the base of the triangle, and h_1 is the height of the triangle. The variable h_2 is the height of the prism.

VOLUME OF A TRAPEZOIDAL PRISM: $V = \frac{1}{2}h_1(b_1 + b_2) \times h_2$

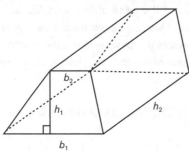

In this case, $B = \frac{1}{2}h_1(b_1 + b_2)$, where h_1 is the height of the trapezoid and b_1 and b_2 are the parallel bases of the trapezoid. The variable h_2 is the height of the prism.

Example:
Given the volume of a rectangular prism is 766.48 mm³, find the height if the length is 6.7 mm and the width is 11 mm.

Use the formula for the volume of a rectangular prism:

$V = lwh$	Substitute in the given information.
$766.48 = 6.7 \times 11 \times h$	Multiply 6.7 times 11.
$766.48 = 73.7 \times h$	Divide 766.48 by 73.7 and include units.
10.4 mm = h	The units are linear (single) because this is a height measurement.

Example:
Find the volume of the trapezoidal prism.

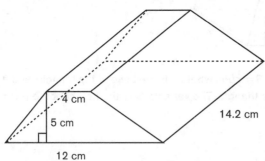

4 cm
5 cm
14.2 cm
12 cm

Use the formula, substituting in the measurements:
$V = \frac{1}{2}h_1(b_1 + b_2) \times h_2$.
$V = \frac{1}{2} \times 5(12 + 4) \times 14.2$.

Evaluate parentheses: $\frac{1}{2} \times 5(16) \times 14.2$.

Multiply left to right to find the volume. The volume is 568 cm³.

CALCULATOR TIPS

If your calculator has the fraction key and parentheses keys, the calculator will most likely perform the correct order of operations. This is especially helpful when finding the surface area or volume of trapezoidal and triangular prisms. Follow the keystrokes below to verify if your calculator follows the order of operations. The example keystrokes are for the previous example. Verify that your calculator shows the correct answer of 568.

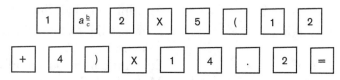

VOLUME OF CYLINDERS AND SPHERES

RULE BOOK

Volume is the area of the base of the solid figure, multiplied by the height of the figure. This can be expressed as $V = Bh$, where V is the volume, B is the area of the base and h is the height of the prism.

VOLUME OF A CYLINDER: $V = \pi r^2 h$

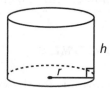

In this case, $B = \pi r^2$, where π is the constant, and r is the radius of the circular base.

VOLUME OF A SPHERE: $V = \frac{4}{3}\pi r^3$

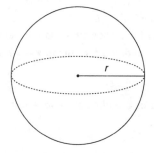

The volume of a sphere is based on the radius of the sphere.

Example:
Find the volume of a cylinder whose base diameter is 14.8 inches and has a height of 18 inches.

First, recognize that while the diameter is given, the radius is needed to calculate volume. Use the formula to find the radius:

$r = \frac{1}{2}d$	Substitute in the given value for diameter.
$r = \frac{1}{2} \times 14.8$	Multiply one-half times 14.8.
$r = 7.4$ in	Now use the formula for the volume of a cylinder.
$V = \pi r^2 h$	Substitute in the given information.
$V = \pi \times 7.4 \times 7.4 \times 18$	Multiply all the number terms together on the right.
$V = 985.68\pi$	Include the cubic units.
$V = 985.68\pi$ in^3	

Example:
Find the volume of a sphere with radius of 12 mm. Use 3.14 for π.

Use the formula $V = \frac{4}{3}\pi r^3$ and substitute in 12 for r; $V = \frac{4}{3}\pi 12^3$. Evaluate the exponent first: $\frac{4}{3}\pi \times 1{,}728$. Multiply to get $2{,}304\pi$ cm^3. Use the value of 3.14 for π to find the volume: 7,234.56 cm^3.

VOLUME OF PYRAMIDS AND CONES

The volume of a pyramid is based on the formula for the volume of a rectangular prism. The volume of a cone is based on the formula for the volume of a cylinder.

RULE BOOK

VOLUME OF A PYRAMID: $V = \frac{1}{3}Bh$
In this case, *B* is the area of the base, and *h* is the height of the pyramid. For a rectangular pyramid, the formula is $V = \frac{1}{3}lwh$.

VOLUME OF A CONE: $V = \frac{1}{3}\pi r^2 h$
In this case, *B* is the area of the circular base, and *h* is the height of the cone.

Example:
The height and the diameter of a cone are both 24 inches. Find the volume of the cone to the nearest hundredth.

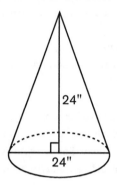

24"

24"

Use the formula for the volume of a cone: $V = \frac{1}{3}\pi r^2 h$. The radius is one-half the measure of the diameter; the radius is 12 inches. Use the π key on the calculator since no value for π is given; $V = \frac{1}{3} \times \pi \times (12)^2 \times 24 = \frac{1}{3} \times \pi \times 144 \times 24 = 1,152 \times \pi$. Multiply using the π key and round to the hundredths place to get 3,6~~19.~~11 in³.

3,617.28

Example:
The volume of the following rectangular pyramid is 508.8 m³. Find the height of the pyramid.

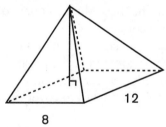

12

8

Represent the height of the pyramid as x. Use the formula for the volume of a pyramid and your knowledge of algebra to find the height; $V = \frac{1}{3} Bh$. The length and width are 8 and 12.

$\frac{1}{3} \times 8 \times 12 \times x = 508.8$	Set up the equation.
$32x = 508.8$	Perform multiplication.
$\frac{32x}{32} = \frac{508.8}{32}$	Divide both sides by 32 and include units.
$x = 15.9$ m	The height has linear units, meters.

EXTRA HELP

If you need extended help in working with surface area and volume, *Geometry Success in 20 Minutes a Day*, published by LearningExpress, has several lessons devoted to this topic: Lesson 14, Lesson 15, and Lesson 16.

There are useful web sites that deal with these topics of geometry. Visit these sites if you feel you need further clarification on these concepts. Each has a unique method of presentation.

1. The website *www.math.com* has extensive lessons on geometry. Once at the site, click on *Geometry*, which you will find on the left under *Select Subject*. From this page, select *Space Figures*. Each topic has a lesson, followed by an interactive quiz. Answers to all quizzes are provided.

2. The website *www.aaamath.com* is another good resource for practice. Once on the home page, click on *Geometry*. You will find this on the right under *Math Topics*. Scroll down to select either *surface area* or *volume* on the right side of the page. The topics are well organized, and there is a brief description of the topic followed by an interactive quiz. Answers are provided.

TIPS AND STRATEGIES

- Familiarize yourself with the common, three-dimensional solids: prism, cylinder, sphere, pyramid, and cone.
- The face of a prism is one of the plane surfaces of the prism.
- The edge of a prism is any one of the segments of the prism.
- A vertex of a prism is where any two edges meet at a point.
- Surface area is the amount of square units it takes to cover or wrap a three-dimensional solid.
- Surface area is measured in square units.
- Volume is a cubic measurement that measures how many cubic units it takes to fill a solid figure.
- Know the volume formulas for the common solid figures.
- The volume of a cone is one-third the volume of a cylinder with the same base and height.
- The volume of a rectangular pyramid is one-third the volume of a rectangular prism with the same base and height.

PRACTICE QUIZ

Use this figure for questions 1 and 2.

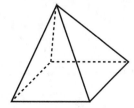

1. How many edges does this solid have?
   a. 9
   b. 12
   c. 6
   d. 8
   e. 5

2. How many faces does this solid have?
   a. 6
   b. 4
   c. 2
   d. 3
   e. 5

3. How many vertices does the following solid figure have?

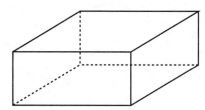

   a. 8
   b. 6
   c. 7
   d. 12
   e. 9

4. Find the surface area of the prism.

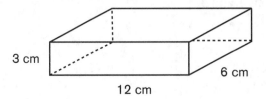

3 cm

6 cm

12 cm

    a. 216 cm²
    b. 252 cm²
    c. 126 cm²
    d. 288 cm²
    e. 432 cm²

5. Find the surface area of a cylinder whose diameter is 16 mm and height is 16 mm.
    a. 384 mm²
    b. 320 mm²
    c. 1,024 mm²
    d. 64π mm²
    e. 384π mm²

6. Find the surface area of the sphere.

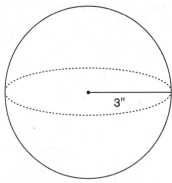

3"

    a. 9π in²
    b. 144π in²
    c. 36π in²
    d. 27π in²
    e. 12π in²

7. Find the volume of the solid.

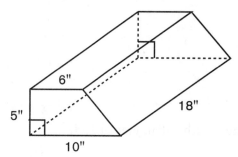

   a. 900 in³
   b. 720 in³
   c. 592 in³
   d. 512 in³
   e. 300 in³

8. If the volume of the triangular prism is 48 cm³, what is the value of the variable x?

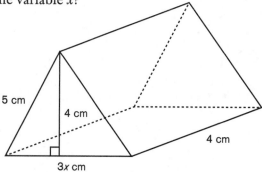

   a. 0.8 cm
   b. 6 cm
   c. 1 cm
   d. 2 cm
   e. 0.5 cm

9. What is the length of a side of a cube whose volume is 27 cm³?
   a. 9 cm
   b. 3 cm
   c. 10 cm
   d. 2.7 cm
   e. 8 cm

10. What is the volume of a cylinder with a height of 100 cm and a radius of 5 m?
   a. $1,000$ m³
   b. 10π m³
   c. $2,500\pi$ m³
   d. $1,000\pi$ m³
   e. 25π m³

11. Find the volume of a sphere with a diameter of 18 ft.
   a. 72 ft³
   b. 729π ft³
   c. 243π ft³
   d. 972π ft³
   e. $1,944\pi$ ft³

12. Find the volume of the pyramid.

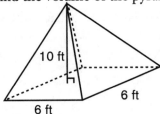

   a. 360 ft³
   b. 360π ft³
   c. 120 ft³
   d. 120π ft³
   e. 180 ft³

13. Find the surface area of the cube.

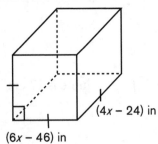

   a. $2,400$ in²
   b. 64 in²
   c. 400 in²
   d. 726 in²
   e. $1,331$ in²

14. Find the surface area of this cylinder where the height is 13 mm.

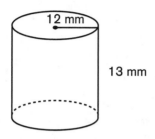

a. 2(144π) + 24 × π × 13 mm²
b. 2(24π) + 2 × π × 13 mm²
c. 144π + 24 × π × 13 mm²
d. 156π mm²
e. 1,872π mm²

15. Find the surface area of the rectangular prism.

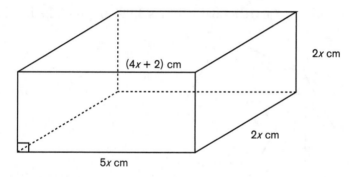

a. 280 cm²
b. 1,600 cm²
c. 160 cm²
d. 240 cm²
e. 192 cm²

16. Find the volume.

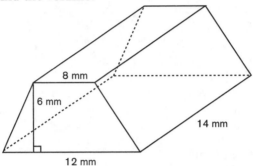

8 mm

6 mm

14 mm

12 mm

   a. 624 mm³
   b. 672 mm³
   c. 840 mm³
   d. 8,064 mm³
   e. 576 mm³

17. If the volume of the triangular prism is 3,990 in³, find the height.

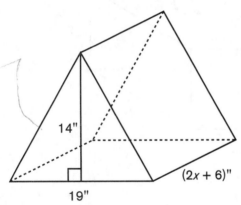

14"

19"

$(2x + 6)$"

   a. 63.4 inches
   b. 30 inches
   c. 28.7 inches
   d. 15 inches
   e. 36 inches

18. The volume of a cube is 512 cm³. What is the value of the variable x, if the length of each side is $2x$ cm long?
   a. 6 cm
   b. 85.3 cm
   c. 64 cm
   d. 4 cm
   e. 8 cm

19. What is the surface area of a sphere with a radius of 6 mm?
  a. 36π mm^2
  b. 144π mm^2
  c. 48π mm^2
  d. 12π mm^2
  e. 288π mm^2

20. Find the volume of the cylinder shown when the height is represented as $16x$ cm.

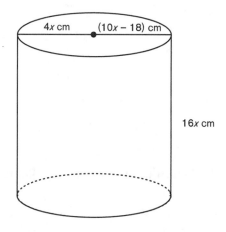

  a. $6{,}912\pi$ cm^3
  b. $1{,}152\pi$ cm^3
  c. 432π cm^3
  d. 96π cm^3
  e. 256π cm^3

21. If the volume of a sphere is $\frac{32\pi}{3}$ m^3, what is the diameter of the sphere?
  a. 16 m
  b. 4 m
  c. 2 m
  d. $\sqrt{8}$ m
  e. 5.66 m

22. The surface area of a cylinder is 182π in^2. What is the height of the cylinder, if the diameter is 14 inches?
  a. 12 inches
  b. 9.5 inches
  c. 1 inch
  d. 11 inches
  e. 6 inches

23. Find the volume of the following cone. The height is 16 mm.

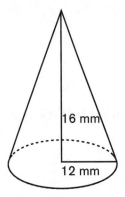

a. 2,304 mm³
b. 768 mm³
c. 768π mm³
d. 2,304π mm³
e. 96π mm³

24. The pyramid has a height of 28 cm. Find the volume to the nearest hundredth.

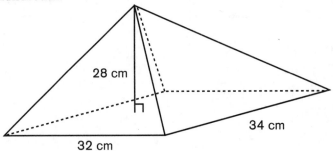

a. 30,464 cm³
b. 10,154.67 cm³
c. 3,384.89 cm³
d. 31.33 cm³
e. 372 cm³

25. Find the volume of the following cone whose height is represented as 3x inches. Use 3.14 for π.

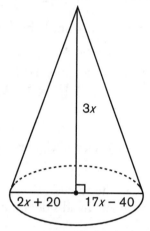

3x

2x + 20   17x − 40

a. 9,408 in³
b. 3,136 in³
c. 29,541.12 in³
d. 9,847.04 in³
e. 703.36 in³

ANSWERS

1. d. There are eight edges, which are the segments of the solid. There are four in the base and four on the sides.

2. e. There are five faces, which are the polygons of the solid. There is one rectangular base and four triangular faces.

3. a. There are eight vertices, the points where the segments meet, on this solid.

4. b. Use the shortcut formula for the surface area of a rectangular prism: $SA = 2(l \times w) + 2(l \times h) + 2(w \times h)$; $SA = 2(12 \times 3) + 2(12 \times 6) + 2(3 \times 6)$. Evaluate parentheses to get $2(36) + 2(72) + 2(18)$. Multiply: $72 + 144 + 36$. Finally, add the measures to get the surface area of 252 square centimeters.

5. e. The diameter is given. First calculate the radius as one-half of the diameter. The radius is 8 mm. Use the formula for the surface area of a cylinder: $SA = 2(\pi r^2) + 2\pi rh$. Substitute in the values: $SA = 2(\pi 8^2) + 2\pi \times 8 \times 16$. Multiply to get $128\pi + 256\pi$. Add these like terms to get 384π mm².

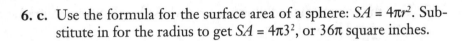

6. c. Use the formula for the surface area of a sphere: $SA = 4\pi r^2$. Substitute in for the radius to get $SA = 4\pi 3^2$, or 36π square inches.

7. b. Volume is the area of the base, a trapezoid, times the height, which is 18 inches. The volume of a trapezoidal prism is $V = \frac{1}{2}b_1$ $(b_1 + b_2) \times b_2$. Substitute in the values from the drawing: $V = \frac{1}{2} \times$ $5(10 + 6) \times 18$. Evaluate parentheses: $V = \frac{1}{2} \times 5(16) \times 18$. Multiply to get 720 cubic inches.

8. d. Use the formula for the volume, $V = \frac{1}{2}bb_1 \times b_2$, and algebra to find the value of the variable x:

$\frac{1}{2} \times 3x \times 4 \times 4 = 48$ Substitute in the values.

$24x = 48$ Simplify the left-hand side: multiply.

$\frac{24x}{24} = \frac{48}{24}$ Divide both sides by 24.

$x = 2$

9. b. The volume of a cube is given by the formula $V = s^3$, where s is the side of a cube. The length of a side of the cube will be the cube root of 27, which is 3, because $3 \times 3 \times 3 = 27$.

10. e. First, note that the units are not consistent. Change 100 cm to 1 meter. Use the formula for the volume of a cylinder: $V = \pi r^2 h$, where r is the radius and h is the height. Substitute in to get $V = \pi$ $\times 5^2 \times 1 = 25\pi$ cubic meters.

11. d. Use the formula for the volume of a sphere: $V = \frac{4}{3}\pi r^3$. The diameter is given. Divide this measure by 2 to find the radius. The radius is 9 feet. Substitute into the formula to get $V = \frac{4}{3}\pi 9^3$. Evaluate the exponent: $V = \frac{4}{3}\pi \times 729$. Multiply to get the volume of 972π cubic feet.

12. c. Use the formula for the volume of a pyramid: $V = \frac{1}{3}lwh$. Substitute in the values from the diagram to get $V = \frac{1}{3} \times 6 \times 6 \times 10$. Multiply to get 120 cubic feet.

13. a. In a cube, all the sides have equal measure. Use this fact and algebra to find the value of the variable x, then find the length of a side.

$6x - 46 = 4x - 24$ Set up an equation.

$6x - 46 - 4x = 4x - 24 - 4x$ Subtract $4x$ from both sides.

$2x - 46 = -24$ Combine like terms.

$2x - 46 + 46 = -24 + 46$ Add 46 to both sides.

$2x = 22$ Combine like terms.

$$\frac{2x}{2} = \frac{22}{2}$$

Divide both sides by 2.

$x = 11$

Use this value to find the length of a side: $6x - 46 = 6(11) - 46 = 66 - 46 = 20$ inches. Surface area is $SA = 6s^2$, or 6 times 20^2. This is 6 times 400, or 2,400 square inches.

14. a. Use the formula for the surface area of a cylinder: $SA = 2(\pi r^2) + 2\pi rh$. Substitute in the values to get $2(\pi \times 12^2) + 2\pi 12 \times 13$. This is equivalent to $2(144\pi) + 24\pi \times 13$ square millimeters.

15. e. In a rectangle, the opposite sides have equal measure. Use this fact and algebra to find the value of the variable x, and then calculate the surface area.

$5x = 4x + 2$ Set up an equation.
$5x - 4x = 4x + 2 - 4x$ Subtract $4x$ from both sides.
$x = 2$ Combine like terms.

Because $x = 2$, the length is 5 times 2, or 10 centimeters, and the width and height are 2 times 2, or 4 centimeters. Use the formula for the surface area of a rectangular prism: $SA = 2(10 \times 4) + 2(10 \times 4) + 2(4 \times 4)$. Evaluate parentheses: $2(40) + 2(40) + 2(16)$. Multiply to get $80 + 80 + 32 = 192$ square centimeters.

16. c. Use the formula for the volume of a trapezoidal prism, $V = \frac{1}{2}h_1(b_1 + b_2) \times h_2$. Substitute in the values from the diagram to get: $V = \frac{1}{2} \times 6(12 + 8) \times 14$. Evaluate parentheses: $V = \frac{1}{2} \times 6(20) \times 14$. Now, multiply to get 840 cubic millimeters.

17. b. Use the formula for the volume of a triangular prism and algebra to find the value of the variable x. Use this value to find the height.

$3,990 = \frac{1}{2}bh_1 \times h_2$

$3,990 = \frac{1}{2} \times 19 \times 14 \times (2x + 6)$ Substitute in the values.

$3,990 = 133(2x + 6)$ Simplify the right side by multiplying.

$3,990 = 266x + 133(6)$ Use the distributive property.

$3,990 = 266x + 798$ Multiply.

$3,990 - 798 = 266x + 798 - 798$ Subtract 798 from both sides.

$3,192 = 266x$ Combine like terms.

$\frac{3,192}{266} = \frac{266x}{266}$ Divide both sides by 266.

$12 = x$

The height is $2x + 6 = 2(12) + 6 = 30$ inches.

18. d. If each side of the cube is $2x$ cm long, use this value and algebra to find the value of the variable; $V = s^3$.

$512 = (2x)^3$	Substitute in the value.
$512 = 8x^3$	Evaluate the exponent.
$\dfrac{512}{8} = \dfrac{8x^3}{8}$	Divide both sides by 8.
$\sqrt[3]{64} = \sqrt[3]{x^3}$	Take the cube root of each side.
$4 = x$	4 times 4 times 4 equals 64.

19. b. Use the formula $SA = 4\pi r^2$. Substitute in the radius: $SA = 4\pi(6^2)$. Evaluate the exponent: $SA = 4\pi(36) = 144\pi$ square millimeters.

20. a. All radii in a circle have the same measure. Use this fact and algebra to find the value of the variable x. Then, use the formula for volume.

$10x - 18 = 4x$	Set up the equation.
$10x - 18 - 4x = 4x - 4x$	Subtract $4x$ from both sides.
$6x - 18 = 0$	Combine like terms.
$6x - 18 + 18 = 0 + 18$	Add 18 to both sides.
$6x = 18$	Combine like terms.
$\dfrac{6x}{6} = \dfrac{18}{6}$	Divide both sides by 6.
$x = 3$	

Because $x = 3$, the radius is $4x$ or $4(3) = 12$ cm, and the height is $16x = 16(3) = 48$ cm. Use the formula $V = \pi r^2 h$, or $V = \pi(12^2)$ times 48. Evaluate the exponent: $144\pi(48) = 6{,}912\pi$ cubic centimeters.

21. b. Use the formula for the volume of a sphere to find the radius. Multiply that value times two to get the diameter: $V = \frac{4}{3}\pi r^3$.

$\dfrac{32\pi}{3} = \dfrac{4}{3}\pi r^3$	Substitute in the values.
$\dfrac{3}{4} \times \dfrac{32\pi}{3} = \dfrac{3}{4} \times \dfrac{4}{3}\pi r^3$	Multiply both sides by the reciprocal of the fraction.
$8\pi = \pi r^3$	Simplify.
$\dfrac{8\pi}{\pi} = \dfrac{\pi r^3}{\pi}$	Divide both sides by π.
$\sqrt[3]{8} = \sqrt[3]{r}$	Take the cube root of each side.
$2 = r$	

Because the radius is 2, the diameter is 4 meters.

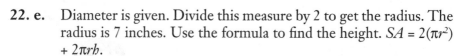

22. e. Diameter is given. Divide this measure by 2 to get the radius. The radius is 7 inches. Use the formula to find the height. $SA = 2(\pi r^2) + 2\pi r h$.

$182\pi = 2(\pi 7^2) + 2\pi \times 7 \times h$	Substitute in the values.
$182\pi = 2(49\pi) + 14\pi(h)$	Evaluate the exponent and multiply.
$182\pi - 98\pi = 98\pi + 14\pi(h) - 98\pi$	Subtract 98π from both sides.
$84\pi = 14\pi(h)$	Combine like terms.
$\frac{84\pi}{14\pi} = \frac{14\pi(h)}{14\pi}$	Divide both sides by 14π.
$6 = h$	The height is 6 inches.

23. c. Use the formula for the volume of a cone: $V = \frac{1}{3}\pi r^2 h$. Substitute in the values given in the diagram: $V = \frac{1}{3}\pi 12^2 \times 16$. Evaluate the exponent: $V = \frac{1}{3}\pi \times 144 \times 16$. Multiply to get 768π cubic millimeters.

24. b. Use the formula for the volume of a pyramid: $V = \frac{1}{3}lwh$. Substitute in the values given in the diagram: $V = \frac{1}{3} \times 32 \times 34 \times 28$. Multiply to get 10,154.67 cubic centimeters, to the nearest hundredth.

25. d. All radii of a circle have the same measure. Use this fact and algebra to solve for the variable x. Use this value to find the needed measures, and then find the volume.

$17x - 40 = 2x + 20$	Set up an equation.
$17x - 40 - 2x = 2x + 20 - 2x$	Subtract $2x$ from both sides.
$15x - 40 = 20$	Combine like terms.
$15x - 40 + 40 = 20 + 40$	Add 40 to both sides.
$15x = 60$	Combine like terms.
$x = 4$	Divide both sides by 15.

The radius is $2x + 20 = 2(4) + 20 = 28$ inches. The height is $3x = 3(4) = 12$ inches.

The formula for the volume of a cone is: $V = \frac{1}{3}\pi r^2 h$. Substitute in the values to get $V = \frac{1}{3} \times 3.14 \times 28^2 \times 12$. Evaluate the exponent: $\frac{1}{3} \times 3.14 \times 784 \times 12$. Multiply to get 9,847.04 cubic inches, to the nearest hundredth.

8

Transformations and Similarity

Previous chapters addressed congruent figures. This chapter deals with figures that undergo transformations, yet are congruent to the original figure. Similar polygons are in proportion to the original figure, and will also be covered. This chapter begins by assessing your understanding of transformations and similarity. Take the benchmark quiz, and assess the knowledge you already possess on this topic. After taking the quiz and reading over the explanations, the lesson will follow to review subjects that you may have forgotten.

BENCHMARK QUIZ

1. What type of transformation does the following picture represent?

   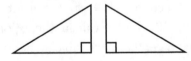

   a. rotation
   b. translation
   c. reflection
   d. dilation
   e. congruence

2. Which of the following letters has rotational symmetry?

I A L T W

   a. the letter I
   b. the letter A
   c. the letter L
   d. the letter T
   e. the letter W

3. Which choice represents all of the symmetries present in the following figure?

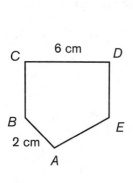

   a. two lines of symmetry only
   b. four lines of symmetry only
   c. two lines of symmetry and rotational symmetry
   d. four lines of symmetry and rotational symmetry
   e. rotational symmetry only

4. Pentagons *ABCDE ~ FGHIJ*. What is the length of $\overline{HI}$?

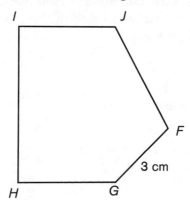

   a. 7 cm
   b. 6 cm
   c. 2 cm
   d. 9 cm
   e. 4 cm

5. If the triangles are similar, what is the scale factor from △QRS to △XYZ?

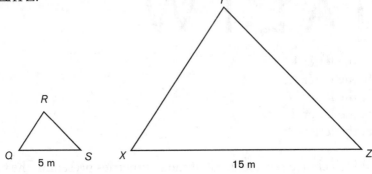

a. 10

b. 3

c. 5

d. $\frac{1}{3}$

e. $\frac{1}{5}$

6. A toy rocket is a model of a real rocket with the scale of 1 : 48. If the toy is 18" long, how long is the real rocket?
  a. 72 inches
  b. 30 inches
  c. 864 feet
  d. 66 inches
  e. 72 feet

7. A man 5 feet tall casts a shadow that is 10 inches long. How tall is the building that casts a shadow that is 14 inches long?
  a. 5 feet, 4 inches
  b. 9 feet
  c. 7 feet
  d. 2 feet
  e. 8 inches

8. The two polygons in the following figure are similar. What is the measure of $\overline{CD}$?

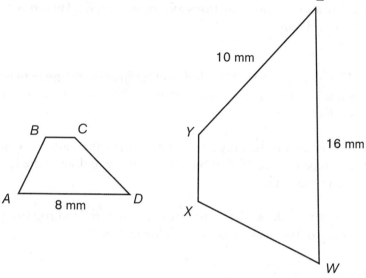

a. 5 mm
b. 20 mm
c. 2 mm
d. 18 mm
e. 12 mm

9. A 10-foot stop sign casts a shadow of 36 inches. How tall is the telephone pole that casts a shadow of 48 inches long?

a. 11 feet
b. $13\frac{1}{3}$ feet
c. 9 feet
d. 22 feet
e. $13\frac{1}{3}$ inches

10. A photographer can blow up a picture. He must use a scale factor or the picture will be distorted. Which dimensions below could be a possible size for an enlargement of a 3" × 5" photograph?

a. 8" × 10"
b. 12" × 15"
c. 9" × 12"
d. 5" × 7"
e. 9" × 15"

BENCHMARK QUIZ SOLUTIONS

Following are the answers to the benchmark quiz. See how much you already know about transformations and similarity. Explanations are provided for all problems.

1. c. The picture represents a flip over an imaginary line between, and equidistant from, the two triangles. This transformation is called a *reflection*.

2. a. The letter I is the only letter choice that has rotational symmetry. It can be rotated 180° about a point at the middle of the letter and map onto itself.

3. d. The figure shown has four lines of symmetry and rotational symmetry of 90°, as shown in the following figure.

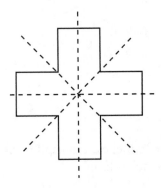

4. d. The pentagons are similar; their sides are in proportion. $\overline{AB}$ corresponds to $\overline{GF}$, and $\overline{CD}$ corresponds to $\overline{HI}$. Set up the proportion:

$$\frac{\text{corresponding side of big pentagon}}{\text{corresponding side of little pentagon}} = \frac{\text{corresponding side of big pentagon}}{\text{corresponding side of little pentagon}}, \text{ and}$$

cross multiply to solve.

$\frac{3}{2} = \frac{\overline{HI}}{6}$	Set up the proportion.
$3 \times 6 = 2 \times \overline{HI}$	Cross multiply.
$18 = 2 \times \overline{HI}$	Multiply on the left side.
$9 = \overline{HI}$	Divide both sides by 2.

5. b. The scale factor is the dilation factor. $\triangle XYZ$ has sides that are three times as big as $\triangle ABC$.

6. e. The ratio of toy rocket to real rocket is 1 to 48. The problem states that the toy rocket is 18 inches long. The real rocket is therefore $18 \times 48 = 864$ inches. This is not an answer choice. Convert 864 inches to feet by dividing by 12, since there are 12 inches in a foot: $\frac{864}{12} = 72$ feet.

7. c. The building and the man, together with their corresponding shadows, form two similar right triangles.

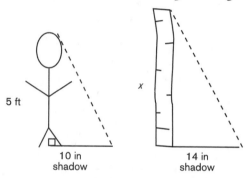

Use a proportion, with the setup of $\frac{\text{shadow of building}}{\text{shadow of man}} = \frac{\text{height of building}}{\text{height of man}}$. Since the units on the shadow are both inches, unit conversion is not needed; the height of the man is in feet, so the height of the building, represented as the variable x, will be in feet.

$\frac{14}{10} = \frac{x}{5}$ Set up the proportion.

$14 \times 5 = 10x$ Cross multiply.

$70 = 10x$ Multiply the left side.

$\frac{70}{10} = \frac{10x}{10}$ Divide both sides by 10.

$7 = x$ The building is 7 feet tall.

8. a. When polygons are similar, the measure of their sides is in proportion. Set up an equation, using:

$\frac{\text{corresponding side of big trapezoid}}{\text{corresponding side of little trapezoid}} = \frac{\text{corresponding side of big trapezoid}}{\text{corresponding side of little trapezoid}}$, and cross multiply to solve. $\overline{AD}$ corresponds to side $\overline{WZ}$, and $\overline{CD}$ corresponds to $\overline{YZ}$:

$$\frac{16}{8} = \frac{10}{\overline{CD}}$$　　　　Set up the proportion.

$16 \times \overline{CD} = 8 \times 10$　　　　Cross multiply.

$16 \times \overline{CD} = 80$　　　　Multiply the right side.

$16 \times \frac{\overline{CD}}{16} = \frac{80}{16}$　　　　Divide both sides by 16.

$\overline{CD} = 5$ mm

9. b. The telephone pole and the stop sign, together with their corresponding shadows, form two similar right triangles.

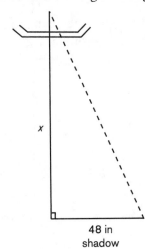
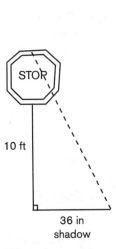

10 ft

36 in
shadow

48 in
shadow

Use a proportion, with the setup of:

$\frac{\text{shadow of pole}}{\text{shadow of stop sign}} = \frac{\text{height of pole}}{\text{height of stop sign}}$. Since the units on the shadow are both inches, unit conversion is not needed; the height of the stop sign is in feet, so the height of the telephone pole, represented as the variable x, will be in feet.

$\frac{48}{36} = \frac{x}{10}$　　　　Set up the proportion.

$48 \times 10 = 36x$　　　　Cross multiply.

$480 = 36x$　　　　Multiply the left side.

$\frac{480}{36} = \frac{36x}{36}$　　　　Divide both sides by 36.

$13.\overline{3} = x$　　　　$.\overline{3}$ is $\frac{1}{3}$.

The telephone pole is $13\frac{1}{3}$ feet tall.

10. e. The original photograph and the enlargement must be similar rectangles. If the rectangles are similar, then each side is enlarged by the same scale factor. The 9" × 15" size is an enlargement with a scale factor of 3. The other dimension choices do not have a consistent scale factor.

BENCHMARK QUIZ RESULTS

If you answered 8–10 questions correctly, feel assured that you have a good foundation in transformations and similarity. Read over the chapter; be sure that you remember all the components of this topic. The sidebars and Tips and Strategies may be especially helpful to you.

If you answered 4–7 questions correctly, you have some understanding of the concepts covered, but you need to carefully study the lessons and sidebars throughout the chapter. Go to the suggested website in the Extra Help sidebar for additional practice. Work through the Practice Quiz at the end of the chapter to check your progress.

If you answered 1–3 questions correctly, you need extended help in understanding this chapter. Take your time as you read through this lesson. Try the examples that are illustrated on a separate sheet of paper and compare your method of solution with that given in the text. Attend to the sidebars and visual aids that will help you grasp the material. Go to the suggested website in the Extra Help sidebar in this chapter, and do extended practice.

JUST IN TIME LESSON— TRANSFORMATIONS AND SIMILARITY

There are many applications of geometry in mathematical problems. Not only will this chapter deal with congruent polygons that are transformed, it will address similar polygons that have the same shape but different size. The topics in this chapter are:

- Transformations
- Symmetry
- Dilations
- Similar Polygons
- Area, Volume, and Similar Polygons
- Applications of Similar Triangles
- Scale

TRANSFORMATIONS

Congruent geometric figures have the same size and shape. All corresponding parts, the sides and angles, have the same measure. Polygons can be moved by a slide, flip, or turn. These movements are called *transformations*.

GLOSSARY

TRANSLATION the image of a geometric figure after a slide in a set direction

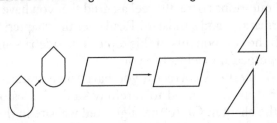

REFLECTION the image of a geometric figure after a flip over a line of symmetry, an imaginary line between and equidistant from the figure and its image

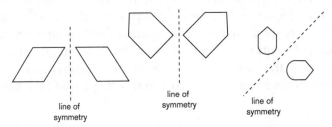

ROTATION the image of a geometric figure after a turn by a set number of degrees around a point

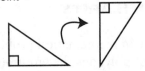

Geometric figures can be transformed over interior lines and points, as shown in the following diagram.

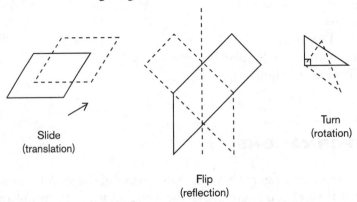

SYMMETRY

When a geometric figure is symmetric, the figure can be reflected over an interior line or rotated from an interior point and it will map onto itself. Two types of symmetry are line symmetry and rotational symmetry. Study a geometric figure; see if you can draw an imaginary line, such that if you folded the figure at this line, the figure would fall on top of itself. There can be none, one, or several lines of symmetry for a figure.

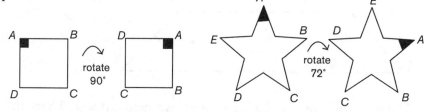

rhombus	parallelogram	star
2 lines of symmetry	No lines of symmetry	5 lines of symmetry

EXTRA HELP

Often, lines of symmetry are readily apparent. If you are unsure, trace the geometric figure and actually fold the traced figure at the line in question to determine if it is in fact a line of symmetry.

Rotational symmetry is present if there is a point in the interior of the figure, usually somewhere in the center, where if you anchor this point and rotate the figure, the figure will fall exactly on top of itself before one complete rotation of 360°.

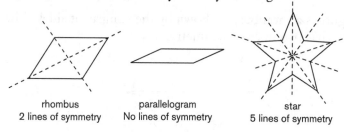

EXTRA HELP

Like lines of symmetry, it is often obvious if a figure has rotational symmetry. If you are unsure, trace the geometric figure. Leave it on top of the original figure, anchor the point of symmetry in question and actually rotate the traced figure.

Example:
How many lines of symmetry does this figure have?

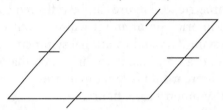

The figure is a rhombus, as shown by the congruent sides. The rhombus has two lines of symmetry:

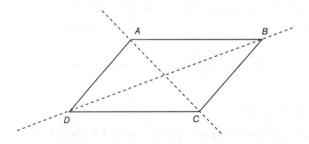

The rhombus also has rotational symmetry. If it is rotated 180° about the point shown, it will map onto itself. Point *A* will be at Point *C*.

Example:
What types of symmetry does this isosceles trapezoid have?

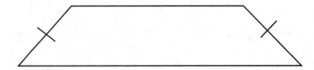

The trapezoid does not have rotational symmetry. It would have to be rotated a full 360° before it would map onto itself. This trapezoid has one line of symmetry:

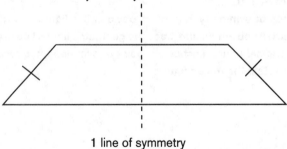

1 line of symmetry

DILATIONS

The previous transformations preserved the size of the figure; the transformed figure was congruent to the original. A fourth transformation, *dilation*, changes the size of a geometric figure, and preserves the shape.

GLOSSARY

DILATION a shrinking or enlarging of a geometric figure that preserves shape but not size

DILATION FACTOR a measure of how the transformed figure has changed size. The dilation factor is a multiplicative operator on the measure of a geometric figure or on the sides of a polygon.

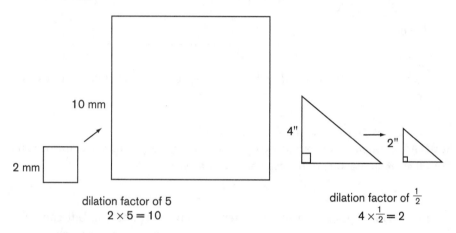

dilation factor of 5
$2 \times 5 = 10$

dilation factor of $\frac{1}{2}$
$4 \times \frac{1}{2} = 2$

Example:
What is the dilation factor from $\triangle PQR$ to $\triangle STU$?

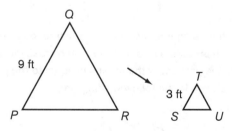

The dilation factor is a fraction, since $\triangle STU$ is smaller than $\triangle PQR$. The factor is $\frac{1}{3}$, because $9 \times \frac{1}{3} = 3$.

SIMILAR POLYGONS

When a polygon is dilated, a similar polygon results. The dilation factor in this case can be called a *scale factor*.

Example:
Rectangle *ABCD* is similar to rectangle *EFGH*. What is the scale factor from *ABCD* to *EFGH*?

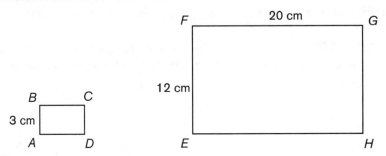

The scale factor is 4; $\overline{EF}$, which corresponds to $\overline{AB}$, is four times as big.

●●●●●● GLOSSARY

SIMILAR polygons have corresponding angles that are congruent, and corresponding sides that are in proportion. The symbol for similarity is ~.

Similar polygons have the same shape (congruent angles), but can have different size. If you are told that two polygons are similar, then their corresponding sides are in proportion. Set up a proportion if the scale factor is not apparent.

 RULE BOOK

Recall that a proportion is an equation in which two ratios are equal. You solve proportional problems by a method called cross multiplication. In a proportion, the product of the means is equal to the product of the extremes, as shown in the following diagram:

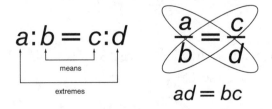

Cross multiplication is a very easy procedure. Follow the example to practice the procedure.

Example:

Solve: $\frac{6}{26} = \frac{n}{52}$; use cross multiplication.

 Show that the product of the means equals the product of the extremes.

$6 \times 52 = 26 \times n$ Multiply 6 and 52.

$312 = 26 \times n$ Divide 312 by 26 to find the missing term.

$n = 12$

The procedure for cross multiplication is straightforward and relatively easy. The challenge in solving word problems using similar polygons is in the set-up of the proportion. Take care to keep all corresponding sides in order. Remember that you are comparing ratios and that the order of the ratio set-up is significant. Choose one method of set-up and use it in all problems. The examples that follow will use the set-up:

$$\frac{\text{corresponding side of big polygon}}{\text{corresponding side of little polygon}} = \frac{\text{corresponding side of big polygon}}{\text{corresponding side of little polygon}},$$

shortened to just: $\frac{big}{little} = \frac{big}{little}$.

Example:

In the following picture, rectangle *ABCD* is similar to rectangle *EFGH*. Find $\overline{FG}$.

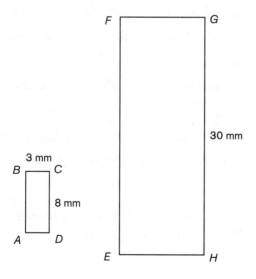

$\overline{DC}$ corresponds to side $\overline{HG}$, and $\overline{BC}$ corresponds to side $\overline{FG}$.

$$\frac{\text{length of } \overline{FG}}{\text{length of } \overline{BC}} = \frac{\text{length of } \overline{GH}}{\text{length of } \overline{CD}}$$ Set up the proportion, knowing that the sides are proportional.

$\frac{\overline{FG}}{3} = \frac{30}{8}$ Substitute in the values of the known sides.

$\overline{FG} \times 8 = 3 \times 30$ Cross multiply.

$\overline{FG} \times 8 = 90$ Multiply on the right side of the equation.

$\overline{FG} = 11.25$ Divide both sides by 8 to get $\overline{FG}$.

Take care and pay attention to the units used in a problem. Sometimes, it is helpful or necessary to convert units before setting up a ratio.

Example:
Rectangle $ABCD \sim EFGH$. What is the measure of $\overline{GH}$?

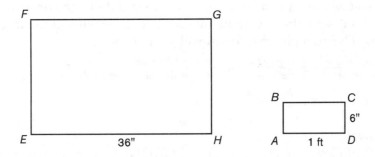

One side of rectangle $ABCD$ is measured in inches, and another side is measured in feet. Convert 1 foot to 12 inches before setting up the proportion. Use the setup $\frac{big}{little} = \frac{big}{little}$, where $\overline{AD}$ corresponds to $\overline{EH}$, and $\overline{CD}$ corresponds to $\overline{GH}$:

$\frac{36}{12} = \frac{\overline{GH}}{6}$ Substitute in the values of the known sides.

$36 \times 6 = 12 \times \overline{GH}$ Cross multiply.

$216 = 12 \times \overline{GH}$ Multiply on the left side of the equation.

$18 = \overline{GH}$ Divide both sides by 12 to get side $\overline{GH}$.

Side $\overline{GH}$ is 18 inches.

Sometimes, the similar polygons are inside each other. Solve this type of similar polygon as you would the others, taking care to use the correct measurements for each polygon.

Example:

The triangles shown in the following diagram are similar. What is the measure of $\overline{CD}$?

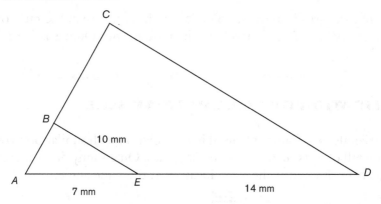

The measure of $\overline{AD}$ is equal to 7 + 14 = 21 mm. This side corresponds to $\overline{AE}$. $\overline{CD}$ corresponds to $\overline{BE}$. Set up a proportion: $\frac{big}{little} = \frac{big}{little}$:

$\frac{21}{7} = \frac{CD}{10}$ Substitute in the values of the known sides.

$21 \times 10 = 7 \times \overline{CD}$ Cross multiply.

$210 = 7 \times \overline{CD}$ Multiply on the left side of the equation.

$30 = \overline{CD}$ Divide both sides by 7 to get $\overline{CD}$.

$\overline{CD}$ is 30 mm.

AREA, VOLUME, AND SIMILAR POLYGONS

Corresponding sides of similar polygons are proportional, and a scale factor establishes the relationship of their measures. Recall from Chapter 6 that area is a multiplication concept—a squaring concept—where two dimensions are multiplied together. If the ratio of the sides of similar polygons is represented as $a : b$, the ratio of the areas will be $a^2 : b^2$. In Chapter 7 you reviewed that volume is a cubing concept, where three measures are multiplied together. If the ratio of the sides of a geometric solid is represented as $a : b$, the ratio of the volumes will be $a^3 : b^3$.

Example:

If the sides of similar rectangles are in the ratio of 2 : 7, what is the ratio of the areas of these rectangles?

The ratio of the areas will be $2^2 : 7^2$, or 4 : 49.

Example:
If the surface area of two similar cubes is in the ratio of 25 : 9, what is the ratio of the volume of these cubes?

Since the area is in the ratio of 25 : 9, the sides are in the ratio of $\sqrt{25} : \sqrt{9}$, or 5 : 3. Therefore, the ratio of the volumes is $5^3 : 3^3$, or 125 : 27.

APPLICATION OF SIMILAR TRIANGLES

Right triangles are commonly used in conjunction with similar polygons to measure tall objects such as trees or flagpoles. On a sunny day, the sun hits all objects positioned at the same location at the same angle.

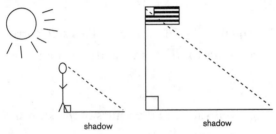

Notice from the above picture that two right triangles are formed, and each triangle has the same angle measures. Because the triangles are similar, the angles are congruent. Therefore, the sides are in proportion. To solve this type of problem, one way is to set up a proportion, such as:

$$\frac{\text{shadow of big}}{\text{shadow of little}} = \frac{\text{height of big}}{\text{height of little}}.$$

Example:
A woman, 68 inches tall, casts a shadow that is 54.4 inches long. If the length of the shadow cast by the flagpole is 192 inches, how tall is the flagpole?

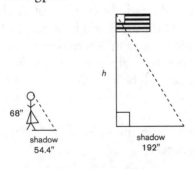

Use the set up: $\dfrac{\text{shadow of big}}{\text{shadow of little}} = \dfrac{\text{height of big}}{\text{height of little}}$

$$\frac{192}{54.4} = \frac{h}{68}$$

Set up the proportion as suggested.

$54.4 \times h = 192 \times 68$ Cross multiply.

$54.4 \times h = 13{,}056$ Divide each side of the equation by 54.4.

$h = 240$ inches, or 20 feet

SCALE

Scale is a special ratio used for models of real-life items, such as model railroads and model airplanes, or scale drawings on blueprints and maps. On model airplanes, you will often find the scale ratio printed on the model as *model : real*. For example, a toy car may have the ratio 1 : 62 printed on the bottom. This is the scale factor of all of the dimensions of the actual toy to the corresponding dimensions of the real car. This scale factor says that the real car is 62 times larger than the toy, since the ratio is 1 : 62.

> *Example:*
> A model locomotive measures 8.7 inches in length. If the scale given is 1 : 16, how long is the real locomotive?
>
> Since the real train is 16 times as big as the model, the real train will be 8.7 times 16, which is 139.2 inches, or 11.6 feet.

On scale drawings, the scale will be a comparison of a small distance unit, like inches, to a large distance unit, like feet. So, a scale on a map could read "3 inches = 10 miles." This means for every 3 inches on the map, it is 10 miles on the actual road. This ratio is $\frac{3}{10}$, but care should be taken to remember that the units do not agree. On a scale drawing, if "1 inch = 10 feet," this does not mean that the real item is 10 times bigger, even though the ratio would be 1 : 10. You would have to convert to like units if you wanted to know how they really compare. Solve scale-drawing problems as you would any type of similarity problem, keeping the units consistent and clear in your answer. Unit conversion is not needed to use the proportion method.

> *Example:*
> A scale drawing of the Statue of Liberty is said to be "$\frac{3}{4}$ inch = 12 feet." How tall is the statue if the scale drawing height is 10 inches? Choose a set-up for the proportion, such as:
>
> $$\frac{\text{drawing dimension}}{\text{real-object dimension}} = \frac{\text{drawing dimension}}{\text{real-object dimension}}$$

$$\frac{\frac{3}{4}}{12} = \frac{10}{h}$$

Set up the proportion, where h stands for height.

$12 \times 10 = \frac{3}{4} \times h$ Cross multiply.

$120 = 0.75 \times h$ Multiply 12 times 10; change $\frac{3}{4}$ to 0.75.

$160 = h$ Divide each side of the equation by 0.75.

According to this drawing, the height of the statue is 160 feet.

EXTRA HELP

For further practice and extended lessons on similarity, refer to *Geometry Success in 20 Minutes a Day:* Lesson 11, Ratio, Proportion, and Similarity. In addition, the website *www.math.com* has helpful mini lessons on similar polygons: 1) Click on *Pre-Algebra* from the leftmost column entitled *Select Subject.* Then, click on the following link, under the title *Ratios and Proportions: Similar Figures.* 2) Click on *Geometry* from the leftmost column entitled *Select Subject.* Then, click on the following link, under the title *Relations and Sizes: Similar Figures.* Another useful website to study transformations is *http://www.utc.edu/ ~cpmawata/.* On the leftmost side, under *Math Cove Projects,* click on *Rigid Transformations.* There are interactive java-based activities to explore transformations.

TIPS AND STRATEGIES

- A translation is a slide of a geometric figure in a set direction.
- A reflection is a flip of a geometric figure over a line.
- A rotation is a turn of a geometric figure around a point.
- Geometric figures can have lines of symmetry and rotational symmetry.
- A dilation is a shrinking or enlarging of a geometric figure by a dilation factor.
- Similar polygons have the same shape, but usually not the same size.
- For similar polygons, corresponding sides are in proportion and corresponding angles are congruent.
- In a proportion, the product of the means equals the product of the extremes.
- Solve proportions using cross multiplication.
- Take care to ensure a proper set-up when solving problems with proportions.
- Scale is a special ratio that compares a model to a real-life object. They are similar geometric figures.

PRACTICE QUIZ

Now that you have studied these lessons, see how much you have learned and reviewed about transformations and similarity.

1. Which of the following statements are true about the figure below?

    a. The figure has exactly one line of symmetry and rotational symmetry.
    b. The figure has exactly five lines of symmetry and no rotational symmetry.
    c. The figure has exactly one line of symmetry and no rotational symmetry.
    d. The figure has exactly five lines of symmetry and rotational symmetry.
    e. The figure has exactly ten lines of symmetry and rotational symmetry.

2. Which figure below shows a single transformation of a translation?

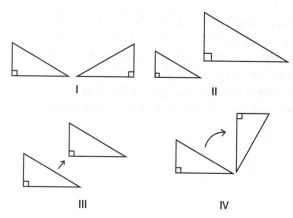

    a. figure I
    b. figure II
    c. figure III
    d. figure IV
    e. none of the above

3. What type of transformation is shown in the following diagram?

    a. rotation
    b. reflection
    c. slide
    d. translation
    e. dilation

4. Which of the following figures has both line symmetry and rotational symmetry?

IMMI OIIO OHIO
        I   II   III

    a. figures I and II
    b. figure II only
    c. figures II and III
    d. figures I and III
    e. all of the above

5. Which of the following statements are NOT ALWAYS true?
    a. A dilation preserves the shape of a geometric figure.
    b. A reflection preserves the shape of a geometric figure.
    c. A translation preserves the size of a geometric figure.
    d. A dilation preserves the size of a geometric figure.
    e. A rotation preserves the shape of a geometric figure.

6. What is the dilation factor from similar triangles *ABC* to *DEF*?

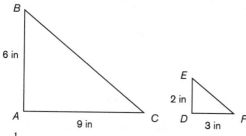

a. $\frac{1}{3}$

b. 3

c. 4

d. $\frac{1}{6}$

e. $\frac{1}{4}$

7. If the dilation factor is 1.5 from rectangle *PQRS* to rectangle *WXYZ*, what is the length of $\overline{XY}$?

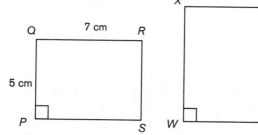

a. 8.5 cm
b. 6.5 cm
c. 3.5 cm
d. 7.5 cm
e. 10.5 cm

8. Quadrilaterals *ABCD* ~ *EFGH*. What is the measure of $\overline{FG}$?

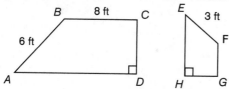

a. 2.25 feet
b. 4 feet
c. 5 feet
d. 3 feet
e. 6 feet

9. The ratio of the sides of two triangles is 3 : 5. What is the ratio of the areas of these triangles?
   a. 6 : 10
   b. 3 : 5
   c. 9 : 15
   d. 15 : 25
   e. 9 : 25

10. A flagpole casts a shadow of 18 feet. A man, 70 inches tall, casts a shadow of 50 inches. How tall is the flagpole?
    a. 25.2 feet
    b. 16.2 feet
    c. 1.35 feet
    d. 196 inches
    e. 16 feet

11. A scale drawing of the Statue of Liberty is 0.6 inches = 10 feet. If the real statue is 160 feet in height, what is the height of the drawing?
    a. 9.6 feet
    b. 960 inches
    c. 9.6 inches
    d. 8 feet
    e. 400 inches

12. An oak tree casts a shadow of 54 inches. The 15-foot telephone pole casts a shadow of 2.25 feet. How tall is the tree?
    a. 30 feet
    b. 440 feet
    c. 400 inches
    d. 78.75 inches
    e. 18.25 feet

13. In the following diagram, $\triangle XYZ \sim \triangle LMN$. What is the length of $\overline{LN}$?

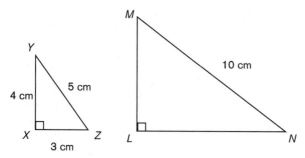

a. 8 cm
b. 9 cm
c. 6 cm
d. 12 cm
e. 7.5 cm

14. A model boat is made to scale. The model has a length of 20 inches and a width of 8 inches. What is the width of the real boat if the length is 36 feet long?
a. 90 feet
b. 7.5 feet
c. 4.4 feet
d. 14.4 feet
e. 53 inches

15. The ratio of the surface area of one face of a cube to the surface area of one face of another cube is 16 : 25. What is the ratio of the volume of the smaller cube to the larger cube?
a. 32 : 50
b. 48 : 75
c. 256 : 625
d. 8 : 12.5
e. 64 : 125

16. The scaled model of a train is 25 cm long. If the real train is 12 meters long, what is the scale factor?
a. 12 : 25
b. 1 : 48
c. 24 : 50
d. 1 : 13
e. 1 : 1,175

17. Given that △QRS ~ △QTU, what is the length of $\overline{QU}$?

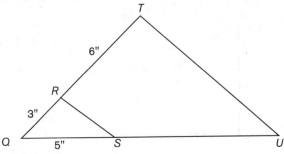

a. 10 inches
b. 15 inches
c. 5 inches
d. 3 inches
e. 8 inches

18. How many lines of symmetry does this parallelogram have?

a. zero lines of symmetry
b. one line of symmetry
c. two lines of symmetry
d. four lines of symmetry
e. six lines of symmetry

19. What type of transformation is shown in the following figure?

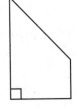

a. translation
b. reflection
c. dilation
d. rotation
e. slide

20. Rectangle *ABCD* is similar to rectangle *EFGH*. What is the length of $\overline{GH}$?

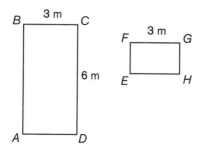

   a. 5 meters
   b. 9 meters
   c. 4.5 meters
   d. 6 meters
   e. 1.5 meters

21. Juan, who is 5.5 feet tall, casts a shadow of 35 inches. How tall is Miguel, who casts a shadow of 42 inches?
   a. 6 feet, 6 inches
   b. 6 feet, 5 inches
   c. 6 feet, 2 inches
   d. 6 feet
   e. 6.6 feet

22. A map shows a scale of 1 cm = 4.5 m. How wide is a lake that measures 5.3 cm on the map?
   a. 238.5 m
   b. 23.85 m
   c. 9.8 m
   d. 980 cm
   e. 1.18 m

23. A drawing is blown up for a billboard. The billboard is 7 feet by 5.6 feet. The drawing must be similar to the billboard. What are possible measurements for the drawing?
   a. 3" × 5"
   b. 8" × 11"
   c. 3" × 4.4"
   d. 8" × 10"
   e. 12" × 20"

24. What is the area of trapezoid *ACDE* if the area of trapezoid *ABGF* is 25, and the trapezoids are similar?

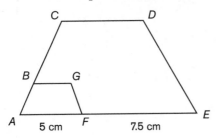

a. 156.25 cm²
b. 62.5 cm²
c. 32.5 cm²
d. 56.25 cm²
e. 27.5 cm²

25. An action figure of King Kong made to scale is 14 inches high. An action figure of Godzilla is made to the same scale. If King Kong was 45 feet tall and Godzilla was 51.75 feet tall, how tall is the Godzilla action figure to the nearest tenth?
a. 20.8 inches
b. 16.1 inches
c. 2.1 feet
d. 1.8 feet
e. 166.3 inches

ANSWERS

1. d. The star has five lines of symmetry and it also has rotational symmetry. The lines of symmetry are shown in the following:

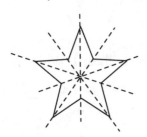

2. c. Figure III shows a translation. Figure I is a reflection, figure II is a dilation, and figure IV is a rotation.

3. b. The drawing shows a trapezoid being reflected, or flipped, over a line between and equidistant from the two images.

4. c. Figures II and III have both line and rotational symmetry of 180°. The lines of symmetry are shown in the following:

5. d. A dilation does not preserve the size of a geometric figure. It usually enlarges or shrinks a figure.

6. a. Triangle *DEF* is smaller, so the dilation factor is a fraction. It is one-third the size of the original triangle *ABC*.

7. e. If the dilation factor is 1.5, multiply the measure of the side that corresponds to $\overline{XY}$, which is $\overline{QR}$, by 1.5; $7 \times 1.5 = 10.5$ cm.

8. b. The quadrilaterals are similar, so their corresponding sides are in proportion. $\overline{FG}$ corresponds to $\overline{BC}$, and $\overline{EF}$ corresponds to $\overline{AB}$. Set up a proportion, using the setup of $\frac{big}{little} = \frac{big}{little}$:

$\frac{8}{\overline{FG}} = \frac{6}{3}$	Set up the proportion.
$8 \times 3 = \overline{FG} \times 6$	Cross multiply.
$24 = \overline{FG} \times 6$	Multiply on the left side of the equation.
$4 = \overline{FG}$	Divide both sides by 6, so $\overline{FG} = 4$ ft.

9. e. If the sides have a ratio of 3 : 5, the areas will have a ratio of these factors squared, that is $3^2 : 5^2 = 9 : 25$.

10. a. The man and the flagpole, together with their corresponding shadows, form similar right triangles.

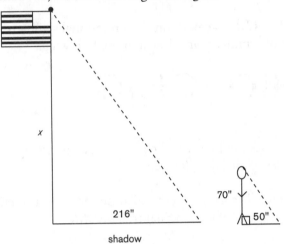

shadow

Use a proportion, with the setup of:

$\frac{shadow\ of\ flagpole}{shadow\ of\ man} = \frac{height\ of\ flagpole}{height\ of\ man}$. Since the units for the man are both inches, convert the flagpole shadow length to inches by multiplying by 12: $18 \times 12 = 216$ inches. Use the variable x to represent the height of the flagpole:

$\frac{216}{50} = \frac{x}{70}$	Set up the proportion.
$216 \times 70 = 50x$	Cross multiply.
$15,120 = 50x$	Multiply on the left side of the equation.
$\frac{15,120}{50} = \frac{50x}{50}$	Divide both sides by 50.
302.4 inches $= x$	

This is not one of the answer choices. Convert this to feet by dividing by 12: $302.4 \div 12 = 25.2$ feet.

11. c. Set up a proportion to solve this problem: $\frac{model}{real} = \frac{model}{real}$. The model is measured in inches, and the real statue is measured in feet. Use the variable x to represent the model height, which will be in inches:

$\frac{0.6}{10} = \frac{x}{160}$	Set up the proportion.
$10x = 0.6 \times 160$	Cross multiply.
$10x = 96$	Multiply on the right side of the equation.
$\frac{10x}{10} = \frac{96}{10}$	Divide both sides by 10.
$x = 9.6$	The height of the drawing is 9.6 inches.

12. a. The tree and the telephone pole, together with their corresponding shadows, form similar right triangles.

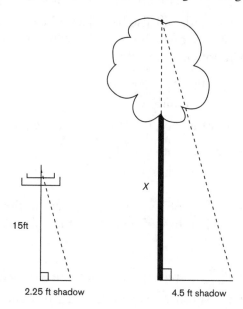

15ft

x

2.25 ft shadow   4.5 ft shadow

Use a proportion; the setup is $\frac{shadow\ of\ tree}{shadow\ of\ pole} = \frac{height\ of\ tree}{height\ of\ pole}$.

Since the units for the telephone pole are both in feet, convert the
tree's shadow length to feet by dividing 54 inches by 12: 54 ÷ 12 =
4.5 feet. Use the variable *x* for tree height:

$\frac{4.5}{2.25} = \frac{x}{15}$ Set up the proportion.

$4.5 \times 15 = 2.25x$ Cross multiply.

$67.5 = 2.25x$ Multiply on the left side of the equation.

$\frac{67.5}{2.25} = \frac{2.25x}{2.25}$ Divide both sides by 2.25.

$x = 30$ The height of the tree is 30 feet.

13. c. The triangles are similar; the sides are in proportion. $\overline{YZ}$ corresponds to $\overline{MN}$ and $\overline{XZ}$ corresponds to $\overline{LN}$. Set up a proportion,
using the setup of $\frac{big}{little} = \frac{big}{little}$.

$\frac{10}{5} = \frac{\overline{LN}}{3}$ Set up the proportion.

$10 \times 3 = 5 \times \overline{LN}$ Cross multiply.

$30 = 5 \times \overline{LN}$ Multiply on the left side of the equation.

$6 = \overline{LN}$ The length of the segment is 6 cm.

14. d. Set up a proportion, comparing width to length. The set up is:
$\frac{width\ of\ model}{length\ of\ model} = \frac{width\ of\ real}{length\ of\ real}$. Note that because the real length of
the boat is in feet, the width will also be in feet. Use the variable x
to represent the width. Because both models' dimensions are in
inches, the units will not affect the set-up.

$\frac{8}{20} = \frac{x}{36}$	Set up the proportion.
$8 \times 36 = 20x$	Cross multiply.
$288 = 20x$	Multiply on the left side of the equation.
$\frac{288}{20} = \frac{20x}{20}$	Divide both sides by 20.
$14.4 = x$	The width is 14.4 feet.

15. e. If the areas are in the ratio of 16 : 25, the sides are in the ratio of
$\sqrt{16}: \sqrt{25}$, or 4 : 5. Therefore, the ratios of the volumes will be
$4^3 : 5^3$, or 64 : 125.

16. b. The scale factor is $\frac{model}{real}$, and the units must be consistent. So,
change the meters to centimeters by multiplying by 100.

$\frac{25}{1,200} = \frac{1}{x}$	Set up the proportion.
$25x = 1,200 \times 1$	Cross multiply.
$25x = 1,200$	Multiply on the right side of the equation.
$\frac{25x}{25} = \frac{1,200}{25}$	Divide both sides by 25.
$x = 48$	The scale factor is 1 : 48.

17. b. The triangles are similar, so set up a proportion using $\frac{big}{little} = \frac{big}{little}$.
The length of $\overline{QT} = 3 + 6 = 9$.

$\frac{9}{3} = \frac{\overline{QU}}{5}$	Set up the proportion.
$9 \times 5 = 3 \times \overline{QU}$	Cross multiply.
$45 = 3 \times \overline{QU}$	Multiply on the left side of the equation.
$15 = \overline{QU}$	The length is 15 inches.

18. a. This parallelogram has no lines of symmetry.

19. d. The figure shows a turn, or a rotation of 90 degrees clockwise.

20. e. The rectangles are similar, so set up a proportion, using $\frac{big}{little} = \frac{big}{little}$. $\overline{CD}$ corresponds to $\overline{FG}$ and $\overline{BC}$ corresponds to $\overline{GH}$.

$\frac{6}{3} = \frac{3}{\overline{GH}}$ Set up the proportion.

$6 \times \overline{QU} = 3 \times 3$ Cross multiply.

$6 \times \overline{QU} = 9$ Multiply on the right side of the equation.

$\overline{QU} = 1.5$ Divide both sides by 6.

21. e. Juan and Miguel, together with their corresponding shadows, form similar right triangles.

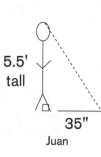 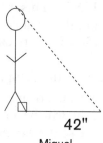

**5.5'
tall**

35"   42"

Juan   Miguel

Use a proportion; the setup is: $\frac{shadow\ of\ Miguel}{shadow\ of\ Juan} = \frac{height\ of\ Miguel}{height\ of\ Juan}$.
Since the units for the shadows are both inches, there is no need to convert units; the units for Miguel's height will be in feet. Use the variable x for Miguel's height:

$\frac{42}{35} = \frac{x}{5.5}$ Set up the proportion.

$42 \times 5.5 = 35x$ Cross multiply.

$231 = 35x$ Multiply on the left side of the equation.

$\frac{231}{35} = \frac{35x}{35}$ Divide both sides by 35.

$6.6 = x$ Miguel is 6.6 feet tall.

22. b. Use a proportion with the setup $\frac{map}{real} = \frac{map}{real}$. The variable x will represent the width of the real lake. Even though the units are different, no conversion is necessary; the answer will be in meters.

$\frac{1}{4.5} = \frac{5.3}{x}$ Set up the proportion.

$x = 4.5 \times 5.3$ Cross multiply.

$x = 23.85$ m   Multiply on the right side of the equation.

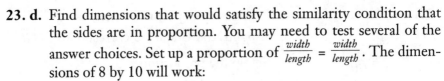
23. d. Find dimensions that would satisfy the similarity condition that the sides are in proportion. You may need to test several of the answer choices. Set up a proportion of $\frac{width}{length} = \frac{width}{length}$. The dimensions of 8 by 10 will work:

$\frac{5.6}{7} = \frac{8}{10}$ Test the proportion.

$5.6 \times 10 = 7 \times 8$ Cross multiply.

$56 = 56$ Multiply on the both sides of the equation. The proportion is true.

24. a. The sides are in a ratio of 5 : 12.5. The areas will be in the ratio of $5^2 : (12.5)^2$, or 25 : 156.25. This is because area is a squaring concept. Since the area of the smaller trapezoid is 25, the area of the larger trapezoid is 156.25 square cm.

25. b. Use a proportion, with the setup: $\frac{model}{real} = \frac{model}{real}$. Since the real heights are both in feet, no unit conversions are needed. The Godzilla model height will be in inches, the same as the King Kong model height. Use the variable x to represent the height of the Godzilla model:

$\frac{14}{45} = \frac{x}{51.75}$ Set up the proportion.

$14 \times 51.75 = 45x$ Cross multiply.

$724.5 = 45x$ Multiply on the left side of the equation.

$\frac{724.5}{45} = \frac{45x}{45}$ Divide both sides by 45.

$16.1 = x$ The Godzilla figure is 16.1 inches tall.

Pythagorean Theorem and Trigonometry

Other chapters in this book have dealt with the most common geometric figure, the triangle. Chapter 4 covered the basic properties and classification of triangles. Chapter 6 reviewed the perimeter and area of triangles. This chapter will review applications that involve the right triangle. Take the following benchmark quiz that starts this chapter to assess the knowledge you already possess about right triangle applications.

BENCHMARK QUIZ

1. In the following right triangle, what is the measure of $\overline{YZ}$, to the nearest hundredth?

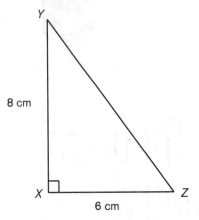

a. 100 cm
b. 5.29 cm
c. 10 cm
d. 3.74 cm
e. 28 cm

2. If the sin ∠*ACB* = 0.5, what is the length of $\overline{AB}$?

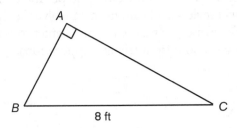

a. 4 feet
b. 5 feet
c. 16 feet
d. 50 feet
e. 10 feet

3. What is the measure of $\overline{PR}$ to the nearest hundredth?

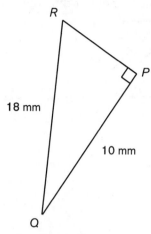

a. 224 mm
b. 14.97 mm
c. 4.05 mm
d. 6.24 mm
e. 2 mm

4. In the following triangle, what is the length of $\overline{XY}$ to the nearest hundredth?

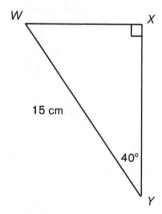

a. 9.64 cm
b. 2.67 cm
c. 300 cm
d. 5.63 cm
e. 11.49 cm

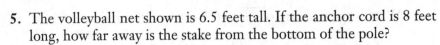

5. The volleyball net shown is 6.5 feet tall. If the anchor cord is 8 feet long, how far away is the stake from the bottom of the pole?

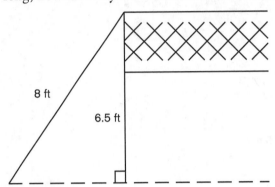

8 ft

6.5 ft

a. 5 feet
b. 1.5 feet
c. 4.67 feet
d. 21.75 feet
e. 16 feet

6. A boat sights a lighthouse at an angle of elevation of 23°. If the lighthouse is 40 feet tall, how far away is the boat from the base of the lighthouse?

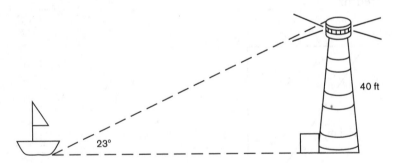

40 ft

23°

a. 63 feet
b. 32.73 feet
c. 15.63 feet
d. 94.23 feet
e. 43.45 feet

7. What is the tangent of ∠*BAC*?

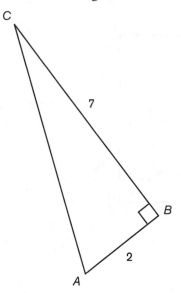

a. $\frac{2}{7}$

b. $\frac{7}{2}$

c. $\sqrt{53}$

d. 7.2

e. 2.7

8. Liz walks 5 blocks north and 3 blocks east to get to school. How much shorter would she walk, to the nearest hundredth of a block, if she could walk the shortest path?
a. 2.17 blocks
b. 4 blocks
c. 5.83 blocks
d. 1 block
e. 8 blocks

9. What is the length of the diagonal of rectangle *ABCD*, to the nearest hundredth?

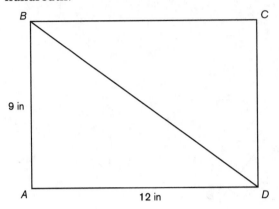

a. 21 inches
b. 31.5 inches
c. 7.94 inches
d. 225 inches
e. 15 inches

10. Given the figure below, what is the length of $\overline{DG}$?

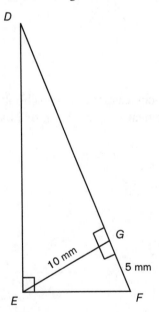

a. 10 mm
b. 5 mm
c. 8.66 mm
d. 20 mm
e. 2.5 mm

BENCHMARK QUIZ SOLUTIONS

1. c. Use the Pythagorean theorem: $a^2 + b^2 = c^2$. Set $a = 6$, $b = 8$, and $c = \overline{YZ}$, to get

$6^2 + 8^2 = c^2$

$36 + 64 = c^2$ Evaluate exponents.

$100 = c^2$ Add like terms on the left.

$\sqrt{100} = \sqrt{c^2}$ Take the square root of both sides.

$10 = c$ $\overline{YZ} = 10$ cm.

2. a. The sin ratio is $\frac{\text{opposite}}{\text{hypotenuse}}$, and $\overline{BC}$ is the hypotenuse, and $\overline{AB}$ is the side opposite to $\angle ACB$. Set up an equation to solve for $\overline{AB}$:

$0.5 = \frac{\overline{AB}}{8}$ Set up the equation.

$4 = \overline{AB}$ Multiply both sides by 8 to get $\overline{AB} = 4$ feet.

3. b. Use the Pythagorean theorem: $a^2 + b^2 = c^2$. Set $a = 10$, $b = \overline{PR}$, and $c = 18$, to get

$10^2 + b^2 = 18^2$

$100 + b^2 = 324$ Evaluate exponents.

$100 + b^2 - 100 = 324 - 100$ Subtract 100 from both sides of the equation.

$b^2 = 224$ Combine like terms.

$\sqrt{b^2} = \sqrt{224}$ Take the square root of both sides.

$b = 14.97$ $\overline{PR} = 14.97$, rounded to the nearest hundredth.

4. e. $\overline{XY}$ is the side adjacent to the angle of 40°, and the side measuring 15 cm is the hypotenuse. Use the cosine ratio to solve for the adjacent side: cosine ratio $= \frac{\text{adjacent}}{\text{hypotenuse}}$.

$\cos 40° = \frac{\overline{XY}}{15}$ Set up the equation.

$(\cos 40°) \times 15 = \overline{XY}$ Multiply both sides by 15.

$11.49 = \overline{XY}$ Perform the cosine function, and multiply to get $\overline{XY}$ rounded to the nearest hundredth.

5. c. The pole and the cord form a right triangle. Use the Pythagorean theorem: $a^2 + b^2 = c^2$. The cord is the hypotenuse of the triangle. The distance needed is b in the formula. Set $a = 6.5$, and $c = 8$, to get

$6.5^2 + b^2 = 8^2$

$42.25 + b^2 = 64$ Evaluate exponents.

$42.25 + b^2 - 42.25 = 64 - 42.25$ Subtract 42.25 from both sides.

$b^2 = 21.75$ Combine like terms.

$\sqrt{b^2} = \sqrt{21.75}$ Take the square root of both sides.

$b = 4.67$ The distance is 4.67 feet, rounded.

6. d. The boat and the lighthouse form a right triangle. The height of the lighthouse is the side opposite to the angle of 23°. The distance from the boat to the lighthouse is the adjacent side to the angle. Use the tangent ratio $= \frac{\text{opposite}}{\text{adjacent}}$. Use the variable x to represent the distance from the boat to the lighthouse.

$\tan 23° = \frac{40}{x}$ Set up the equation.

$(\tan 23°)x = 40$ Multiply both sides by x.

$94.23 = x$ Divide both sides by the tangent of 23°, to find the distance, rounded to the nearest hundredth.

7. b. The tangent ratio is $\frac{\text{opposite}}{\text{adjacent}}$. The side opposite the angle is 7 and the side adjacent to the angle is 2, or $\frac{7}{2}$.

8. a. Draw a picture of the situation.

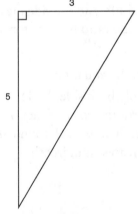

Her walk forms a right triangle, and the hypotenuse would be the shortest path. Use the Pythagorean theorem: $a^2 + b^2 = c^2$. Set $a = 5$, and $b = 3$, to get

$5^2 + 3^2 = c^2$

$25 + 9 = c^2$ Evaluate exponents.

$34 = c^2$ Combine like terms.

$\sqrt{34} = \sqrt{c^2}$ Take the square root of both sides.

$5.83 = c$ The shortest distance is 5.83
blocks, rounded.

Her walk was 5 + 3 = 8 blocks. The shortcut is 5.83 blocks. She would thus walk 8 – 5.83 = 2.17 blocks shorter to get to school.

9. e. Use the Pythagorean theorem: $a^2 + b^2 = c^2$. The diagonal is the hypotenuse of a right triangle formed by two of the sides. Set $a = 9$, and $b = 12$, to get

$9^2 + 12^2 = c^2$

$81 + 144 = c^2$ Evaluate exponents.

$225 = c^2$ Combine like terms.

$\sqrt{225} = \sqrt{c^2}$ Take the square root of both sides.

$15 = c$ The length of the diagonal, an exact
measurement.

10. d. The figure shows an altitude drawn to the hypotenuse of a large right triangle, forming two smaller right triangles. The three right triangles are similar. Break the triangles apart to see the relationship of the sides. $\overline{EG}$ is a shared side to both.

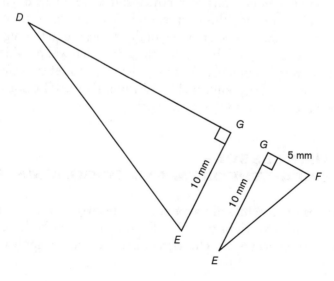

Set up the sides of the smaller triangles in a proportion:
$\dfrac{\overline{DG}}{\overline{EG}} = \dfrac{\overline{EG}}{\overline{FG}}.$

$\dfrac{\overline{DG}}{10} = \dfrac{10}{5}$ Set up the proportion.

$\overline{DG} \times 5 = 10 \times 10$ Cross multiply.

$\overline{DG} \times 5 = 100$ Multiply on the right side of the equation.

$\overline{DG} = 20$ Divide both sides by 5.

BENCHMARK QUIZ RESULTS:

If you answered 8–10 questions correctly, you have a good understanding of right triangle relationships and applications. Perhaps the questions you answered incorrectly deal with one specific area in this chapter. Read over the chapter, concentrating on those areas of weakness. Proceed to the Practice Quiz at the end of the chapter and try to improve your score.

If you answered 4–7 questions correctly, there are several areas you need to review. Carefully read through the lesson in this chapter for review and skill building. Work carefully through the examples and pay attention to the sidebars referring you to definitions, hints, and shortcuts. Get additional practice on geometry by visiting the suggested websites and working through the Practice Quiz at the end of the chapter.

If you answered 1–3 questions correctly, spend some time studying this chapter. By carefully reading this chapter and concentrating on the visual aids, you will gain a better understanding of this topic. Go to the suggested websites in the Extra Help sidebar in this chapter, which will provide additional help and extended practice. You may also want to refer to *Geometry Success in 20 Minutes a Day*, published by LearningExpress. Lessons 8 and 20 in the book are devoted to these concepts.

JUST IN TIME LESSON—
PYTHAGOREAN THEOREM AND TRIGONOMETRY

The Pythagorean theorem and trigonometry are two very common and practical applications for right triangles. Follow the lesson in this chapter to review the basic concepts of right triangles, which you can apply in word problems.

The topics in this chapter are:

- Similar Right Triangles
- Identifying the Sides of a Right Triangle
- The Pythagorean Theorem
- Special Right Triangles
- Applications of the Pythagorean Theorem
- Identifying the Opposite and Adjacent Sides of a Right Triangle
- The Trigonometric Ratios
- Applications of Trigonometry

SIMILAR RIGHT TRIANGLES

Chapter 8 described similar polygons. A special case of similarity arises with right triangles.

 RULE BOOK

When an altitude is drawn from the hypotenuse of a right triangle, two smaller right triangles emerge. All three triangles are similar to one another.

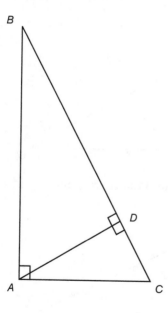

When this type of problem is encountered, it is helpful to break the three triangles apart as shown:

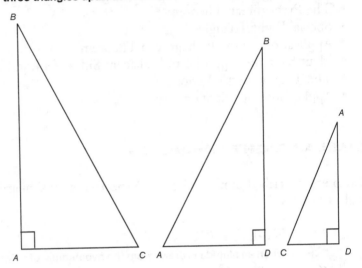

Example:

In the following diagram, what is the length of $\overline{DC}$?

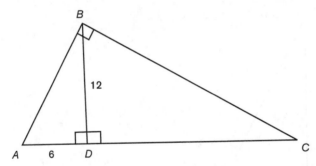

Separate the triangles, and find the corresponding sides:

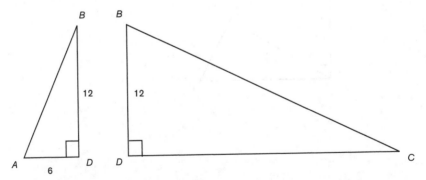

$\overline{DC}$ on the larger of the two triangles corresponds to $\overline{BD}$ on the smaller of the two triangles. $\overline{BD}$ on the larger of the two triangles corresponds to $\overline{AD}$ on the smaller of the two triangles.

Set up a proportion, using $\frac{big}{little} = \frac{big}{little}$, $\frac{\overline{BD}}{\overline{AD}} = \frac{\overline{DC}}{\overline{BD}}$:

$\frac{12}{6} = \frac{x}{12}$ Set up the proportion, using the variable x for the unknown.

$12 \times 12 = 6x$ Cross multiply.

$144 = 6x$ Multiply on the left side of the equation.

$\frac{144}{6} = \frac{6x}{6}$ Divide both sides by 6.

$24 = x$ $\overline{DC}$ is 24 inches.

IDENTIFYING THE SIDES OF A RIGHT TRIANGLE

Right triangles are special triangles used for measuring objects that would otherwise be difficult to measure directly, such as the height of a tall tree. In a right triangle, the base and one side are perpendicular.

GLOSSARY

HYPOTENUSE of a right triangle is the side of the right triangle that is opposite the right angle

LEGS of a right triangle are the two sides of the right triangle that make up the right angle

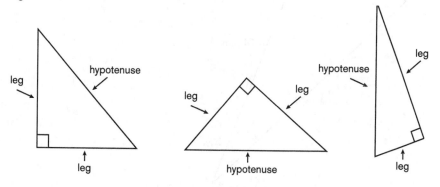

In right triangles, there is a special relationship between the hypotenuse and the legs of the triangle. This relationship is always true and it is known as the Pythagorean theorem.

THE PYTHAGOREAN THEOREM

RULE BOOK

The Pythagorean theorem states that in all right triangles, the sum of the squares of the two legs is equal to the square of the hypotenuse; leg² + leg² = hypotenuse².

The converse of the Pythagorean theorem is also true: In a triangle, the sum of the squares of the legs is equal to the square of the hypotenuse if and only if the triangle is a right triangle.

SHORTCUT

You can remember the Pythagorean theorem as the well-known formula: $a^2 + b^2 = c^2$, where a and b are the two legs of the right triangle, and c is the hypotenuse.

Special note: Be careful! There is nothing special about the letters a, b, and c. A test question could be "tricky" and could call one of the legs c.

Example:
Find m to the nearest hundredth.

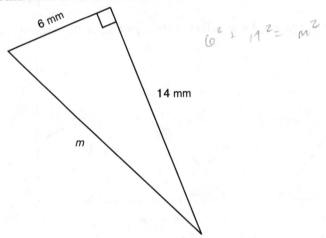

Use the Pythagorean theorem $a^2 + b^2 = c^2$, where a and b are 6 mm and 14 mm and c is the unknown, m. Note that it does not matter whether you set $a = 6$ and $b = 14$, or $a = 14$ and $b = 6$, due to the commutative property of addition.

$a^2 + b^2 = c^2$ Substitute in the given lengths.
$6^2 + 14^2 = m^2$ Evaluate the exponents, left to right.
$36 + 196 = m^2$ Perform addition.
$232 = m^2$ Take the square root of 232 to find m.
$\sqrt{232} = m$ The value of m will be approximate, to the
                                        nearest hundredth and include the units.

15.23 mm $= m$

CALCULATOR TIPS

Use the $\boxed{x^2}$ key on the calculator to find the square of a number. To calculate 14^2 on the calculator, enter:

$\boxed{1}\ \boxed{4}\ \boxed{x^2}\ \boxed{=}$

Use the

$\boxed{\sqrt{}}$ or $\boxed{\overset{\sqrt{}}{x^2}}$

2nd function

key to find the square root of a number. Depending on your specific calculator, this key is used either before or after you enter the radicand. To find $\sqrt{289}$, enter

$\boxed{\sqrt{}}\ \boxed{2}\ \boxed{8}\ \boxed{9}\ \boxed{=}$

or

$\boxed{2}\ \boxed{8}\ \boxed{9}\ \boxed{\sqrt{}}\ \boxed{=}$

Example:
Find the value of y in the following diagram:

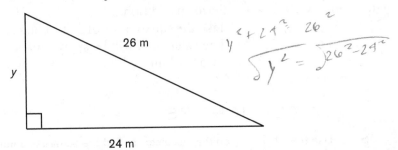

$$y^2 + 24^2 = 26^2$$
$$y^2 = \sqrt{26^2 - 24^2}$$

This is a right triangle, where the hypotenuse is 26 m and one of the legs is 24 m. This is a multiple of the common Pythagorean triple, 5, 12, 13, so the Pythagorean triple is 10, 24, 36 by multiplying each length by two. The unknown side, *y*, is therefore 10 m.

SHORTCUT

There are three sets of Pythagorean triples that appear over and over again in math test problems. Knowing these three common triples will save you valuable time in working problems of this type.

	a	b	c
One set is:	3	4	5
and multiples thereof:	6	8	10
	9	12	15
	12	16	20
Another set is:	5	12	13
and multiples thereof:	10	24	26
	15	36	39
The third set is:	8	15	17
	16	30	34

Memorize these sets: {3, 4, 5}, {5, 12, 13} and {8, 15, 17}. If a right triangle problem is given and two of the three numbers in one set appear (or multiples of the two numbers), you can avoid all the substituting and calculating and save precious test time.

SPECIAL RIGHT TRIANGLES

Some special right triangles have side relationships that are helpful to memorize.

SHORTCUT

In a 30°–60°–90° triangle, the hypotenuse is twice the length of the shorter side, and the longer side length is the length of the shorter side, multiplied by the square root of three:

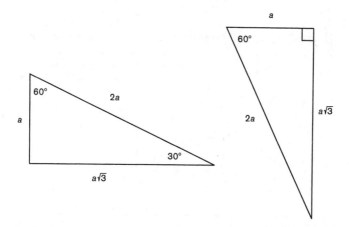

In an isosceles right triangle, the length of the hypotenuse is the length of a leg, multiplied by the square root of two:

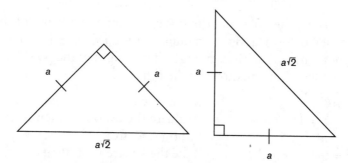

The altitude of an equilateral triangle forms two smaller congruent triangles that are 30°–60°–90° triangles.

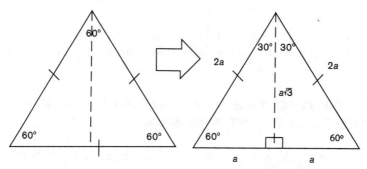

APPLICATIONS OF THE PYTHAGOREAN THEOREM

The Pythagorean theorem is often embedded in word problems, where a right triangle is formed from the given situation.

Example:
A volleyball net is staked to the ground as shown in the diagram:

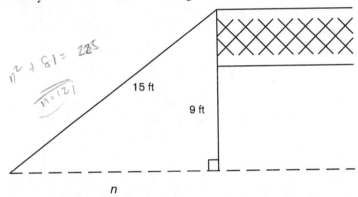

The cord on the stake is 15 feet, and the pole is 9 feet tall. How far from the bottom of the pole is the stake (value n in the diagram)?

The cord, the pole, and the distance from the bottom of the pole to the stake form a right triangle. The cord in this diagram is the hypotenuse of the right triangle. The height of the pole, 9 feet, is one of the legs. The unknown is the other leg.

$a^2 + b^2 = c^2$	Substitute in the given lengths.
$9^2 + b^2 = 15^2$	Evaluate the exponents, left to right.
$81 + b^2 = 225$	Subtract 81 from 225.
$b^2 = 144$	Take the square root of 144 to find b.
$b = \sqrt{144}$	The value of b is 12; include units in the answer.

$b = 12$ ft

Note that this problem could have been solved using the shortcut, as 9, 12, 15 is a multiple of the common Pythagorean triple of 3, 4, 5.

IDENTIFYING THE OPPOSITE AND ADJACENT SIDES OF A RIGHT TRIANGLE

The legs of a right triangle can be defined in reference to one of the acute angles in the triangle.

GLOSSARY

ADJACENT side to an acute angle, in a right triangle, is the side of the triangle that is a side of the angle, and not the hypotenuse

OPPOSITE side to an acute angle, in a right triangle, is the side of the triangle that is across from the angle in question

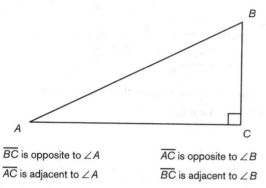

$\overline{BC}$ is opposite to $\angle A$ $\overline{AC}$ is opposite to $\angle B$

$\overline{AC}$ is adjacent to $\angle A$ $\overline{BC}$ is adjacent to $\angle B$

Note that the side that is adjacent to one of the acute angles is opposite to the other acute angle.

THE TRIGONOMETRIC RATIOS

The word *trigonometry* means, "measure of right triangles." Common comparisons, or ratios, are made between the sides of right triangles. These are known as the *trigonometric ratios*, and these ratios reference one of the acute angles in a right triangle.

 GLOSSARY

SINE (SIN) of an angle is the ratio that compares the side opposite the angle, in a right triangle, to the hypotenuse of the triangle

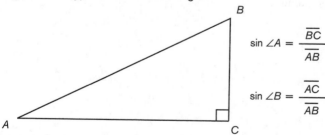

$$\sin \angle A = \frac{\overline{BC}}{\overline{AB}}$$

$$\sin \angle B = \frac{\overline{AC}}{\overline{AB}}$$

COSINE (COS) of an angle is the ratio that compares the side adjacent the angle, in a right triangle, to the hypotenuse of the triangle

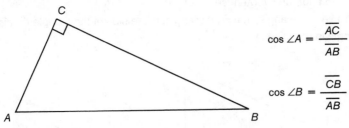

$$\cos \angle A = \frac{\overline{AC}}{\overline{AB}}$$

$$\cos \angle B = \frac{\overline{CB}}{\overline{AB}}$$

TANGENT (TAN) of an angle is the ratio that compares the side opposite the angle, in a right triangle, to the side adjacent to the angle

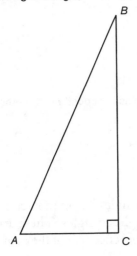

$$\tan \angle A = \frac{\overline{BC}}{\overline{AC}}$$

$$\tan \angle B = \frac{\overline{AC}}{\overline{BC}}$$

SHORTCUT

You may recall from your school days, the mnemonic device to remember the trigonometric ratios: SOH–CAH–TOA, or S$\frac{O}{H}$–C$\frac{A}{H}$–T$\frac{O}{A}$, where $S =$ sin, $C =$ cos, $T =$ tan, $O =$ opposite, $A =$ adjacent, and $H =$ hypotenuse.

Example:
In the following right triangle, what is the tan $\angle C$?

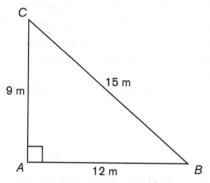

The side opposite to $\angle C$ is 12 m and the side adjacent to $\angle C$ is 9 m.
Use the tangent ratio of $\frac{\text{opposite}}{\text{adjacent}}$, to get $\frac{12 \text{ m}}{9 \text{ m}} = \frac{12}{9}$, or alternately $\frac{4}{3}$ in lowest terms.

Sometimes, the ratios are given as a fraction, and other times they are given as a decimal.

Example:
If the cos $\angle Q = 0.5$ in the following triangle, what is the length of $\overline{PQ}$?

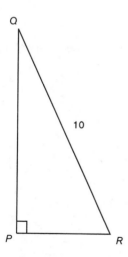

The cosine ratio is $\frac{\text{adjacent}}{\text{hypotenuse}}$, so set up an equation:

$\frac{\overline{PQ}}{10} = 0.5$ Substitute in the known values.

$\overline{PQ} = 5$ Multiply both sides by 10.

CALCULATOR TIPS

You will need a scientific calculator to find the trigonometric ratios. Most often, tests will present angles in degrees. First, ensure that your calculator is in degree mode. The method of changing the angle mode will vary for each individual calculator, so check your manual. Try the examples below, and others in this chapter, to be sure your calculator is in the correct mode.

To find a trigonometric ratio from the calculator, use the following keys:

Your calculator will return the ratio as a decimal. Some calculators require the angle measure followed by the specific *trig* key and others require the *trig* key first, followed by the angle measure. Check your calculator with the example below to see which type you have:

Example:
Find the sin of 30°.

   or

Your calculator should show 0.5.

In some problems, the trigonometric ratio is given for a right triangle and the actual angle measure is requested. Use your calculator to find these angle measures.

Example:
Given the following right triangle, what is the measure of ∠XYZ, to the nearest tenth of a degree?

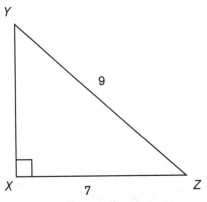

The side opposite to ∠XYZ has a length of 7 and the hypotenuse has a length of 9. Use the sine ratio, $\frac{opposite}{hypotenuse}$, and the sin^{-1} key on the calculator.

$$\sin \angle XYZ = \frac{7}{9}$$ Use the $\sin^{-1}$ key.

If your calculator has fraction keys, enter the fraction when using the $\sin^{-1}$ key. Otherwise, use 0.7777 as an approximation for the given fraction.

The angle is $\angle XYZ = 51.1°$, rounded to the nearest tenth of a degree.

CALCULATOR TIPS

If the trigonometric ratio is given, and the angle is required, use the following calculator keys, which are typically the second function above the regular trigonometry keys:

$\sin^{-1}$ $\cos^{-1}$ $\tan^{-1}$

[sin]   [cos]   [tan]

Example:

If the tangent of an angle is 1.0, what is the value of the angle?
Use the calculator and enter these strokes:

$\tan^{-1}$
[2nd] [tan] [1] [=]

Your calculator should show 45.

If the trigonometric ratio is rounded, the angle degree measure may also have to be rounded, as in the following example:

Example: If the sin of an angle is approximately 0.866, what is the value of the angle?
Use the calculator and enter these strokes:

$\sin^{-1}$

Your calculator gives an angle measure of 59.99708907, which would round to 60°. Caution: Do not round trigonometric values until the end of the calculations. When a problem asks for a rounded answer, use the calculator to keep accuracy throughout the calculations, and round only the final answer. If your calculator does not have parentheses or fraction keys, use at least four decimal places in the approximation of any trigonometric ratio.

APPLICATIONS OF TRIGONOMETRY

Trigonometry is often embedded in word problems where a right triangle is formed from the given situation. A common application uses *an angle of elevation* or an *angle of depression*. A surveyor uses a tool to calculate the height of an object, such as a tree. He stands a set distance from the tree, and then measures the angle of elevation to the top of the tree:

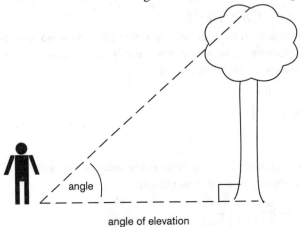

angle of elevation

Note that this situation is in essence a right triangle trigonometry problem, where the height of the tree is the side opposite to the angle of elevation.

An airplane pilot, knowing his height above the ground, can determine the length of the horizontal distance between the plane and the edge of the runway by measuring the angle of depression:

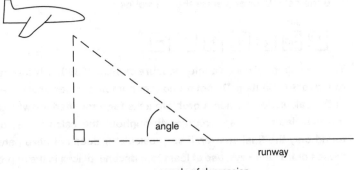

angle of depression

Again, this situation is a right triangle trigonometry problem. Note that the angle of depression is the same angle used in the angle of elevation of the preceding problem.

Example:

A fisherman wants to know how far his boat is from shore. On the edge of the shore is a lighthouse, which stands 50 feet in height. The fisherman sights the top of the lighthouse at an angle of 25°. How far is the boat from shore, to the nearest tenth?

Draw a picture of the situation:

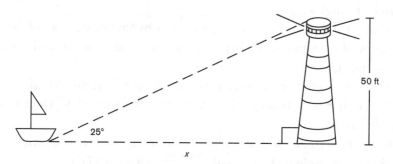

The height of the lighthouse is the side opposite the angle of 25°. The distance from the boat to the shore is the side adjacent to the 25° angle. Use the tangent ratio, $\frac{\text{opposite}}{\text{adjacent}}$, to find the distance, represented as x:

$\tan 25° = \dfrac{50}{x}$	Set up the equation with the given information.
$(\tan 25°)x = 50$	Multiply both sides by x.
$\dfrac{(\tan 25°)x}{(\tan 25°)} = \dfrac{50}{(\tan 25°)}$	Divide both sides by the tangent of 25°.
$x = 107.2$ feet	Round to the nearest tenth.

EXTRA HELP

If you need extended help in working with geometry, refer to the book *Geometry Success in 20 Minutes a Day*, published by LearningExpress. There are useful web sites that deal with the Pythagorean theorem and trigonometry.

1. The website *http://library.thinkquest.org/20991/geo/stri.html* is a tutorial on the concepts covered in this chapter.

2. The website *www.math.com* is another source of help. Once at the site, click on *Geometry*, which you will find on the left under *Select Subject*. From this page, select *The Pythagorean Theorem and right triangle facts*, under *Relations and Sizes*. Each topic has a lesson, followed by an interactive quiz. Answers to all quizzes are provided.

TIPS AND STRATEGIES

- When an altitude to a right triangle is drawn, two smaller right triangles are formed, and the three triangles are similar.
- Familiarize yourself with the parts of a right triangle.
- The Pythagorean theorem is used to make measurements with right triangles.
- The Pythagorean theorem is leg^2 + leg^2 = hypotenuse^2, or $a^2 + b^2 = c^2$.
- Know the common Pythagorean triples and how to find their multiples.
- The trigonometric ratios compare the sides of a right triangle.
- Be familiar with triangle sides that are opposite and adjacent to a given angle.
- Know how to use your calculator with the trigonometric ratios.
- Sine of an angle is the ratio of $\frac{\text{opposite}}{\text{hypotenuse}}$, $S\frac{O}{H}$, or *SOH*.
- Cosine of an angle is the ratio of $\frac{\text{adjacent}}{\text{hypotenuse}}$, $C\frac{A}{H}$, or *CAH*.
- Tangent of an angle is the ratio of $\frac{\text{opposite}}{\text{adjacent}}$, $T\frac{O}{A}$, or *TOA*.
- For word problems, draw a picture, and identify the right triangle formed.

PRACTICE QUIZ

1. What is the value of *n* in the following right triangle, to the nearest hundredth?

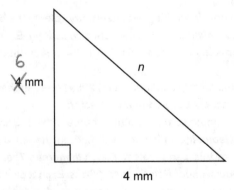

a. 4.47 mm
b. 7.21 mm
c. 5 mm
d. 10 mm
e. 3.16 mm

2. What is the sin ∠A?

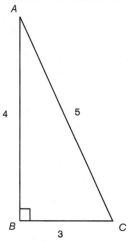

a. $\dfrac{3}{5}$

b. $\dfrac{4}{5}$

c. $\dfrac{3}{4}$

d. $\dfrac{4}{3}$

e. 45°

3. What is the length of $\overline{MN}$ in the following triangle, to the nearest hundredth?

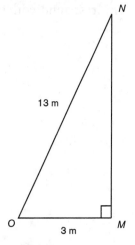

a. 13.93 m
b. 2.45 m
c. 8 m
d. 144 m
e. 12 m

4. What is the tangent of ∠*P*?

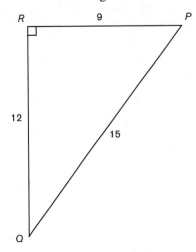

  a. $\frac{9}{12}$

  b. $\frac{12}{15}$

  c. $\frac{12}{9}$

  d. $\frac{9}{15}$

  e. 53°

5. Lara walks 7 blocks west and then 2 blocks south to get to the stadium. How much shorter is the walk, to the nearest hundredth block, if she could walk a straight line?
  a. 7.28 blocks
  b. 1.72 blocks
  c. 4.5 blocks
  d. 4.24 blocks
  e. 2 blocks

6. When it is 5,000 feet above the ground, an airplane sights the 10-foot tall control tower at an angle of depression of 50°. What is the horizontal distance, to the nearest foot, from the tower to the plane?

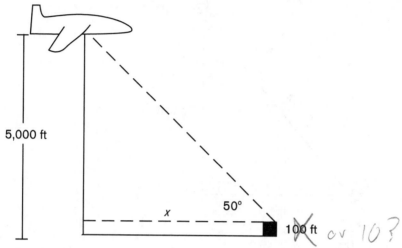

a. 4,195 feet
b. 6,527 feet
c. 7,779 feet
d. 4,187 feet
e. 5,959 feet

7. What is the length of the diagonal, to the nearest hundredth, of a square whose side is 12.5 cm long?
   a. 25 cm
   b. 7.07 cm
   c. 17.68 cm
   d. 12.5 cm
   e. 156.25 cm

8. Katy rides the chair lift on the ski slope. The lift is 1,200 feet long, and rises 1,000 feet vertically. What is the angle, $\angle B$, that the slope makes with the horizontal, to the nearest degree?

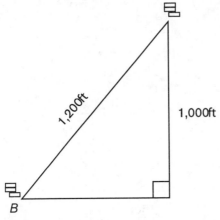

a. 56°
b. 200°
c. .83°
d. 34°
e. 12°

9. What is the length of $\overline{HI}$, in the following right triangle, to the nearest tenth?

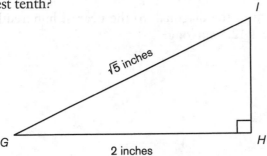

a. 11.5 inches
b. 21 inches
c. 3 inches
d. 1 inch
e. 4.6 inches

10. If the sin $\angle T$ = 0.8, what is the length of $\overline{ST}$?

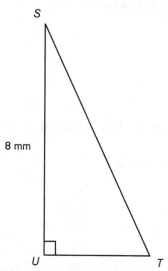

8 mm

a. 8.04 mm
b. 8.09 mm
c. 4.2 mm
d. 6 mm
e. 10 mm

11. In $\triangle WXV$, if $\overline{XV}$ = 7 feet, what is the length of $\overline{WX}$?

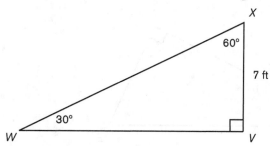

60°

7 ft

30°

a. $7\sqrt{3}$ feet
b. $7\sqrt{2}$ feet
c. 14 feet
d. 3.5 feet
e. 7 feet

12. $\triangle QRS$ is an isosceles right triangle. What is the length of the hypotenuse if the leg of the triangle is $\sqrt{8}$ inches long?
    a. 4 inches
    b. $8\sqrt{2}$ inches
    c. 16 inches
    d. 2 inches
    e. $\sqrt{2}$ inches

13. A conveyor belt at the shipping company is constructed as shown:

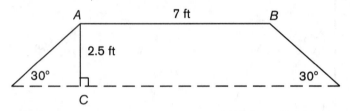

How long is the full length of the conveyor belt, to the nearest foot?
    a. 5 ft
    b. 10 ft
    c. 12 ft
    d. 17 ft
    e. 13 ft

14. If the sin $\angle DEF = \frac{4}{5}$, what is the cos $\angle DFE$?

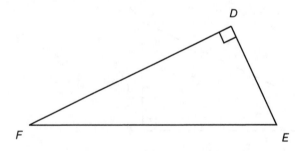

    a. $\frac{4}{5}$

    b. $\frac{3}{5}$

    c. $\frac{3}{4}$

    d. $\frac{5}{4}$

    e. $\frac{5}{3}$

15. What is the height of an equilateral triangle whose side is 20 mm long?

   a. $20\sqrt{3}$ mm

   b. $10\sqrt{3}$ mm

   c. $10\sqrt{2}$ mm

   d. 7.75 mm

   e. 10 mm

16. What is the measure of ∠XYZ?

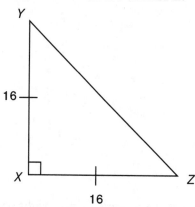

   a. 22.63°

   b. 90°

   c. .8°

   d. 60°

   e. 45°

17. What is the measure of a side of a square whose diagonal is 10 inches long, to the nearest hundredth?

   a. 5 inches

   b. 7.07 inches

   c. 14.14 inches

   d. 3.16 inches

   e. 4.47 inches

18. In the following triangle, what is the length of $\overline{AB}$ to the nearest tenth?

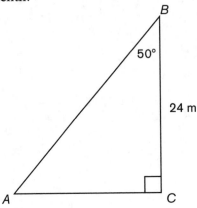

a. 31.3 m
b. 20.1 m
c. 15.4 m
d. 37.3 m
e. 18.4 m

19. Which of the given side measures for a triangle forms a right triangle?
a. 3, 4, and 6
b. 9, 12, and 13
c. 6, 7, and 12
d. 12, 15, and 17
e. 15, 20, and 25

20. What is the measure of ∠*JKL* to the nearest degree?

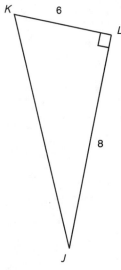

   a. 53°
   b. 37°
   c. .49°
   d. 41°
   e. 13°

21. What is the length of $\overline{AC}$, to the nearest hundredth?

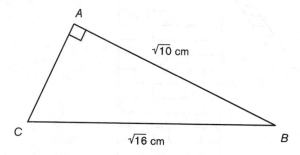

   a. 6 cm
   b. 12 cm
   c. 3.46 cm
   d. 2.45 cm
   e. 12.49 cm

22. If the cos $\angle LMN$ is $\frac{6}{10}$, what is the sin $\angle LMN$?

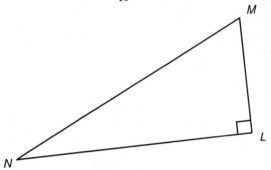

a. $\frac{6}{10}$

b. $\frac{10}{6}$

c. $\frac{8}{10}$

d. $\frac{10}{8}$

e. $\frac{6}{8}$

23. A 17-foot ladder rests on a building as shown:

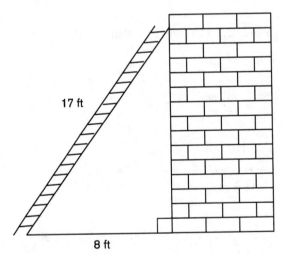

17 ft

8 ft

If the base of the ladder is 8 feet away from the building, how high up the building does the ladder touch?

a. 19 ft

b. 18 ft

c. 15 ft

d. $\sqrt{98}$ ft

e. $\sqrt{353}$ ft

24. What is the length of x in the following diagram?

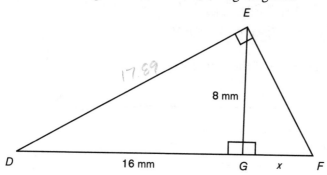

a. 1 mm
b. 8 mm
c. 2 mm
d. 6 mm
e. 4 mm

25. A surveyor sights the top of a tree at an angle of elevation of 70°. If he is 15 feet away from the bottom of the tree, how tall is the tree to the nearest foot?
a. 14 ft
b. 41 ft
c. 5 ft
d. 68 ft
e. 10 ft

ANSWERS

1. b. Use the Pythagorean theorem: $a^2 + b^2 = c^2$. Set $a = 6$, $b = 4$, and $c = n$, to get

$6^2 + 4^2 = n^2$
$36 + 16 = n^2$ Evaluate exponents.
$52 = n^2$ Add like terms on the left.
$\sqrt{52} = \sqrt{n^2}$ Take the square root of both sides.
$7.21 = n$ The value is 7.21 to the nearest hundredth.

2. a. The sine ratio is $\frac{\text{opposite}}{\text{hypotenuse}}$. The side opposite to $\angle A$ is 3 and the hypotenuse is 5.

3. e. You can use the Pythagorean theorem to find the length of the missing leg, or you can recognize that this is a common Pythagorean triple, namely the 5–12–13 triple.

4. c. The tangent ratio is $\frac{\text{opposite}}{\text{adjacent}}$. The side opposite to $\angle P$ is 12 and and the side adjacent to $\angle P$ is 9.

5. b. Draw a picture of the situation.

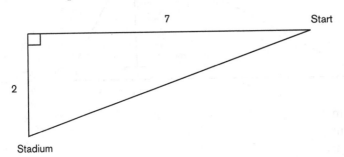

Her walk forms a right triangle, and the hypotenuse would be the shortest path. The shortest distance between two points is a straight line. Use the Pythagorean theorem: $a^2 + b^2 = c^2$. Set $a = 7$, and $b = 2$, to get

$7^2 + 2^2 = c^2$	
$49 + 4 = c^2$	Evaluate exponents.
$53 = c^2$	Combine like terms.
$\sqrt{53} = \sqrt{c^2}$	Take the square root of both sides
$7.28 = c$	The shortest distance is 7.28, rounded.

Her walk was $7 + 2 = 9$ blocks. The shortcut is 7.28 blocks. She would thus walk $9 - 7.28 = 1.72$ blocks shorter to get to the stadium.

6. d. From the diagram, the side opposite the angle of 50° is $5,000 - 10$ or 4,990. The horizontal distance from the tower to the plane is the side adjacent to the angle. Use the tangent ratio, which is $\frac{\text{opposite}}{\text{adjacent}}$. Use the variable x to represent the horizontal distance.

$\tan 50° = \frac{4,990}{x}$	Set up the equation.
$(\tan 50°)x = 4,990$	Multiply both sides by x.
$x = 4,187$	Divide both sides by the tan 50° to get the distance, rounded to the nearest foot.

7. c. Draw a picture:

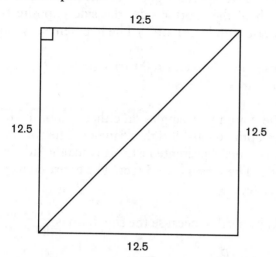

The diagonal is the hypotenuse of an isosceles triangle, whose sides are 12.5. This is a special triangle, so the hypotenuse is $12.5 \times \sqrt{2}$ = 17.68 cm, to the nearest hundredth.

8. a. From the diagram, the chair lift length is the hypotenuse and the vertical distance is the side opposite to $\angle B$. Use the sine ratio, which is $\frac{\text{opposite}}{\text{hypotenuse}}$.

$\sin \angle B = \frac{1,000}{1,200}$ Set up the equation.

$\angle B = 56$ Use the $\sin^{-1}$ key to find the angle, rounded to the nearest degree.

9. d. Use the Pythagorean theorem: $a^2 + b^2 = c^2$. Set $a = 2$, and $c = \sqrt{5}$, to get

$2^2 + b^2 = (\sqrt{5})^2$
$4 + b^2 = 5$ Evaluate exponents.
$4 + b^2 - 4 = 5 - 4$ Subtract 4 from both sides.
$b^2 = 1$ Take the square root of both sides.
$b = 1$ $\overline{HI} = 1$ inch

10. e. The sine ratio is $\frac{\text{opposite}}{\text{hypotenuse}}$. The side opposite is 8 and the hypotenuse is $\overline{ST}$. Set up an equation:

$0.8 = \frac{8}{\overline{ST}}$ Set up an equation.
$0.8 \times \overline{ST} = 8$ Multiply both sides by $\overline{ST}$.
$\overline{ST} = 10$ Divide both sides by 0.8.

11. c. Δ*WXV* is a 30–60–90 special right triangle. The hypotenuse is twice the length of the shorter side, the side opposite the 30° angle. (In this case with length of 7.) The hypotenuse is thus 14.

12. a. The hypotenuse of an isosceles right triangle is √2 × leg of the triangle; √2 × √8 = √16 = 4.

13. d. Notice that there are two triangles on either end of the belt system, which are both 30–60–90 right triangles. The conveyor belt section on either end, represented as *x*, is twice the side opposite to the 30° angle. The variable *x* = 5 feet. The entire conveyor belt is 5 + 7 + 5 = 17 feet.

14. a. Draw in the known sides, because the sine ratio is $\frac{\text{opposite}}{\text{hypotenuse}}$:

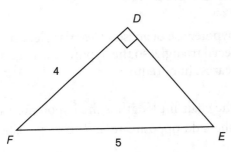

The cosine ratio is $\frac{\text{adjacent}}{\text{hypotenuse}}$, and the adjacent side to ∠*DFE* is 4. The cos ∠*DFE* is $\frac{4}{5}$.

15. b. Draw a picture of the triangle, with the height shown:

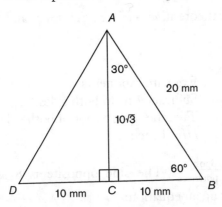

The height is also an altitude, so it forms two 30–60–90 right triangles. The side opposite to the 30° angle, $\overline{DC}$, is one half the hypotenuse, 20 mm. $\overline{DC}$ = 10 mm. The side opposite to the 60° angle, the altitude, is 10√3 mm long.

16. e. In the diagram, the sides opposite and adjacent to the angle are shown. Use the tangent ratio to find the angle measure. Tangent is $\frac{\text{opposite}}{\text{adjacent}}$.

$\tan \angle XYZ = \frac{16}{16}$ Set up the equation.

$\tan \angle XYZ = 1$

$\angle XYZ = 45°$ Use the $\tan^{-1}$ key to find the angle.

This problem could have been easily solved without trigonometry. This is an isosceles right triangle, so the acute angle is 45°.

17. b. Draw the square, with the diagonal. The diagonal is the hypotenuse of an isosceles right triangle. Use the Pythagorean theorem, $a^2 + b^2 = c^2$. Set $a = b$, and $c = 10$, to get

$a^2 + a^2 = 10^2$

$2a^2 = 100$ Combine like terms.

$a^2 = 50$ Divide both sides by 2.

$\sqrt{a^2} = \sqrt{50}$ Take the square root of both sides.

$a = 7.07$ The side of the square is 7.07 to the nearest hundredth.

18. d. From the diagram, the side of length 24 is adjacent to the angle, and $\overline{AB}$ is the hypotenuse of the right triangle. Use the cosine ratio, that is $\frac{\text{adjacent}}{\text{hypotenuse}}$.

$\cos 50° = \frac{24}{AB}$ Set up the equation.

$\cos 50° \times \overline{AB} = 24$ Multiply both sides by $\overline{AB}$.

$\overline{AB} = 37.3$ Divide both sides by the $\cos 50°$, and round.

19. e. Test the pairs by using the Pythagorean theorem. If the theorem holds, the triangle is a right triangle. If it does not hold, the triangle is NOT a right triangle. Test each choice:

Choice **a**: $3^2 + 4^2 = 6^2$, $9 + 16 = 36$ is NOT true. Choice **b**: $9^2 + 12^2 = 13^2$, $81 + 144 = 169$ is NOT true. Choice **c**: $6^2 + 7^2 = 12^2$, $36 + 49 = 144$ is NOT true. Choice **d**: $12^2 + 15^2 = 17^2$, $144 + 225 = 289$ is NOT true. Choice **e**: $15^2 + 20^2 = 25^2$, $225 + 400 = 625$ is a true statement. Choice **e** is the right triangle. Time could have been saved if it had been recognized that 15, 20, and 25 is a multiple of the common {3, 4, 5} triple.

20. a. From the diagram, the opposite and adjacent side measures are given for the angle. Use the tangent ratio, which is $\frac{\text{opposite}}{\text{adjacent}}$.

$\tan \angle JKL = \frac{8}{6}$ Set up the equation.

$\angle JKL = 53°$ Use the $\tan^{-1}$ key, and round to the nearest degree.

21. d. Use the Pythagorean theorem, $a^2 + b^2 = c^2$. Set $a = \sqrt{10}$, and $c = \sqrt{16} = 4$, to get

$\sqrt{10}^2 + b^2 = 4^2$

$10 + b^2 = 16$ Evaluate exponents.

$10 + b^2 - 10 = 16 - 10$ Subtract 10 from both sides.

$b^2 = 6$ Combine like terms.

$\sqrt{b2} = \sqrt{6}$ Take the square root of both sides.

$b = 2.45$ Rounded to the nearest hundredth.

22. c. Draw in the sides based on the cosine ratio, $\frac{\text{adjacent}}{\text{hypotenuse}}$, as shown:

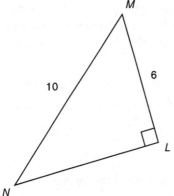

The sides indicate a multiple of the common Pythagorean triple, {3, 4, 5}, that is {6, 8, 10}. The sine ratio of the given angle is $\frac{\text{opposite}}{\text{hypotenuse}}$, or $\frac{8}{10}$.

23. c. From the picture, this is a right triangle problem, where the hypotenuse is 17 and one of the sides is 8. You can use the Pythagorean theorem, or you can recognize the common Pythagorean triple of 8–15–17. The height where the ladder rests against the building is 15 feet.

24. e. The figure shows an altitude drawn to the hypotenuse of a large right triangle, forming two smaller right triangles. The three right triangles are similar. Break the triangles apart to see the relationship of the sides. $\overline{EG}$ is a shared side to both.

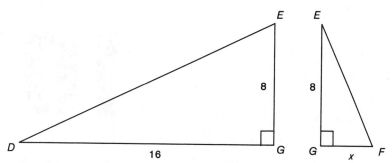

Set up the sides of the smaller triangles in a proportion: $\frac{DG}{EG} = \frac{EG}{FG}$

$\frac{16}{8} = \frac{8}{FG}$ Set up the proportion.

$16 \times \overline{FG} = 8 \times 8$ Cross multiply.

$16 \times \overline{FG} = 64$ Multiply on the right side.

$\overline{FG} = 4$ Divide both sides by 16.

25. b. Draw a picture of the situation:

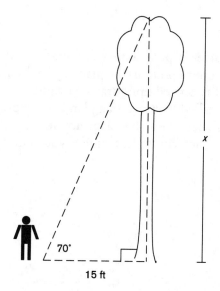

The height of the tree is represented as x, and is opposite to the angle. The side measuring 15 feet is the side adjacent to the angle. Use the tangent ratio, which is $\frac{\text{opposite}}{\text{adjacent}}$.

$\tan 70° = \frac{x}{15}$ Set up the equation.

$\tan 70° \times 15 = x$ Multiply both sides by 15.

$41 = x$ Use the tan key, and then multiply and round.

Coordinate Geometry

In the 1600s, Rene Descartes, a philosopher and mathematician, developed a method of positioning a point in the plane by its distances, x and y, from the intersection of two fixed lines drawn at right angles in the plane. This plane came to be called the Cartesian plane, or simply, the coordinate plane. This chapter reviews how this plane applies to geometry. Take the benchmark quiz to determine how much you remember about coordinate geometry.

BENCHMARK QUIZ

1. What are the coordinates of point R on the graph below?

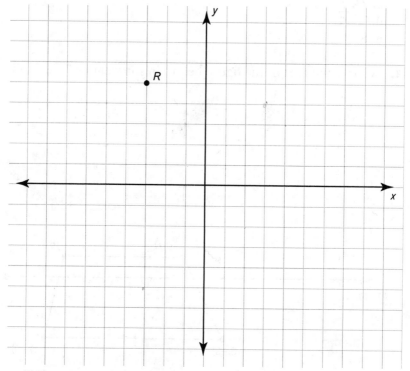

   a. (5,3)
   b. (−5,3)
   c. (5,−3)
   d. (3,5)
   e. (−3,5)

2. One endpoint of a segment is (3,4). What is the other endpoint, if the midpoint of the segment is (−1,3)?
   a. (2,7)
   b. (−3,12)
   c. (1,3.5)
   d. (−5,2)
   e. (2,−5)

3. What is the distance between points $(-3,-5)$ and $(2,-2)$?
   a. $2\sqrt{6}$
   b. $5\sqrt{2}$
   c. $2\sqrt{5}$
   d. $\sqrt{34}$
   e. 8

4. What is the area of the figure below, in square units?

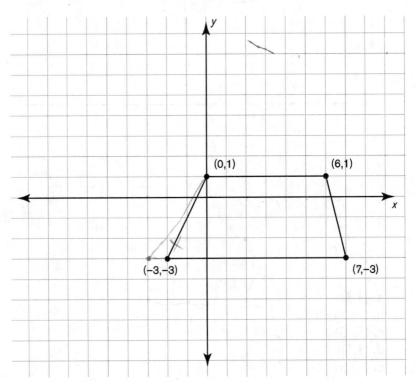

   a. 40 sq. units
   b. 24 sq. units
   c. 32 sq. units
   d. 64 sq. units
   e. 8 sq. units

5. What is the equation of the line on the graph below?

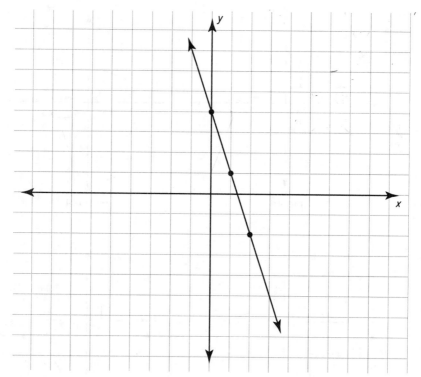

  a. $y = 3x + 4$

  b. $y = -3x + 4$

  c. $y = 4x + 3$

  d. $y = 4x - 3$

  e. $y = \frac{1}{3}x + 4$

6. What is the equation of a line that is perpendicular to the line $y = -4x + 5$?

  a. $y = 4x + \frac{1}{5}$

  b. $y = -\frac{1}{4}x + 5$

  c. $y = 4x + 5$

  d. $y = -5x - 5$

  e. $y = \frac{1}{4}x - 3$

7. Which graph shows a reflection of △*ABC* over the line *y* = *x*?

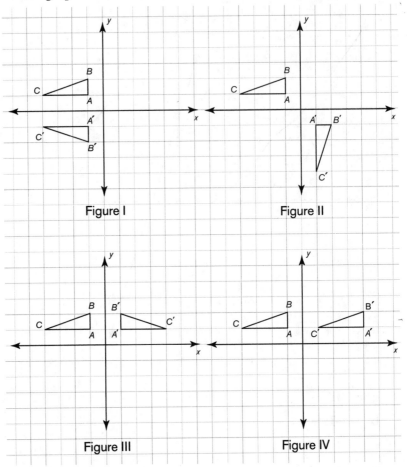

a. Figure I
b. Figure II
c. Figure III
d. Figure IV
e. none of the above

8. What is the solution to the system of equations shown below?

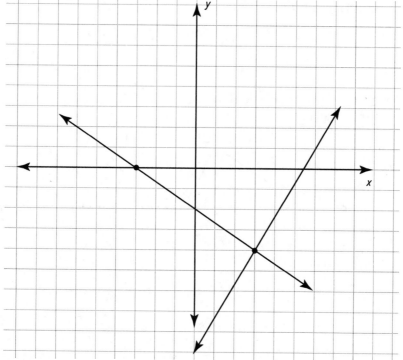

a. (3,–4)
b. (0,–2)
c. (0,–7)
d. (–4,3)
e. (–3,0)

9. What is the transformation shown in the graph below, from $\triangle ABC$ to its image $\triangle A'B'C'$?

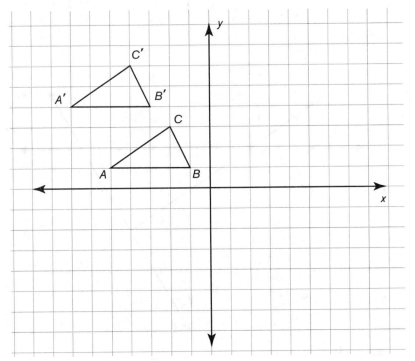

a. $T_{(-2,3)}$
b. $R_{P,-90°}$
c. $r_{y\text{-axis}}$
d. $r_{x\text{-axis}}$
e. $T_{(2,-3)}$

10. What is the equation of a line that is parallel to $y = 7x - 15$?

a. $y = \frac{1}{7}x - 15$
b. $y = 15x + 7$
c. $y = 7x - 8$
d. $y = -7x - 8$
e. $y = -7x + 3$

BENCHMARK QUIZ SOLUTIONS

1. e. Point R is three units to the left of the origin, so the x-coordinate is -3 and it is 5 units above the origin, so the y-coordinate is 5. The coordinates are $(-3,5)$.

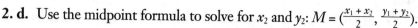
2. d. Use the midpoint formula to solve for x_2 and y_2: $M = (\frac{x_1 + x_2}{2}, \frac{y_1 + y_2}{2})$.

$3 + \frac{x_2}{2} = -1$	Use the midpoint formula.
$3 + x_2 = -2$	Multiply both sides by 2.
$x_2 = -2 - 3$	Isolate the x and y.
$x_2 = -5$	Combine like terms.

$\frac{4 + y_2}{2} = 3$

$4 + y_2 = 6$

$y_2 = 6 - 4$

$y_2 = 2$

The coordinates are $(-5, 2)$

3. d. Use the distance formula: $D = \sqrt{(x_2 - x_1)^2 + (y_2 - y_1)^2}$

$D = \sqrt{(-3 - 2)^2 + (-5 - -2)^2}$, or $D = \sqrt{(-5)^2 + (-3)^2}$, or $D = \sqrt{25 + 9} = \sqrt{34}$.

4. c. Use the formula for the area of a trapezoid: $A = \frac{1}{2}h(b_1 + b_2)$. Count the units for b_1 (the base), b_2 (the other base) and h (the height). Base 1, b_1, is $7 - -3 = 10$ units long. Base 2, b_2, is $6 - 0 = 6$ units long. The height, h, is $1 - -3 = 4$ units long. Substitute in these values to get: $A = \frac{1}{2} \times 4(10 + 6)$, or $A = \frac{1}{2} \times 64 = 32$ square units.

5. b. The graphed line crosses the y-axis at $(0,4)$, so the y-intercept is 4. The slope can be calculated from the points $(0,4)$ and $(1,1)$, using $\frac{\text{rise}}{\text{run}} = \frac{y_2 - y_1}{x_2 - x_1} = \frac{4 - 1}{0 - 1} = \frac{3}{-1} = -3$. The equation is $y = -3x + 4$.

6. e. The slope of a perpendicular line will have a slope that is the negative reciprocal of the given equation. The slope is the coefficient before the variable x when the equation is in the form $y = mx + b$. The slope of the given equation is -4. The negative reciprocal is $\frac{1}{4}$. Choice **e** is the only choice with this slope.

7. b. Figure II shows a reflection over the line $y = x$, which is shown below:

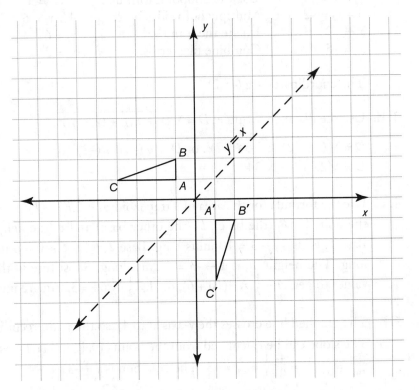

Figure I is a reflection over the x-axis. Figure III is a reflection over the y-axis. Figure IV is a translation.

8. a. The solution to the system is the coordinates of the point of intersection. This point is three units to the right of the origin, so the x-coordinate is 3, and four units below the origin, so the y-coordinate is –4. The coordinates are (3,–4).

9. a. This transformation is a slide in which each point slides two units to the left (negative direction) and three units up (positive direction). This is $T_{(-2,3)}$.

10. c. The equation of a line parallel to the given equation will have the same slope. The only equation that has the same slope, which is 7, is choice **c.** When an equation is in the form $y = mx + b$, such as these, the slope is the coefficient of the x variable.

BENCHMARK QUIZ RESULTS:

If you answered 8–10 questions correctly, you possess a solid foundation of knowledge about coordinate geometry. Pay attention to the chapter sidebars to be sure you have a full understanding of the various aspects of this topic. Work through the Practice Quiz at the end of the chapter to ensure your success.

If you answered 4–7 questions correctly, carefully read through the lesson in this chapter to review the material and build your confidence. Work carefully through the examples and pay attention to the sidebars that refer you to definitions, hints, and shortcuts. Get additional practice on geometry by visiting the suggested websites and solving the Practice Quiz at the end of the chapter.

If you answered 1–3 questions correctly, concentrate your efforts on Chapter 10 topics. First, carefully read this chapter including the sidebars and visual aids that will improve your comprehension. Go to the suggested websites in the Extra Help sidebar in this chapter, which will provide extended practice. You may also want to refer to *Geometry Success in 20 Minutes a Day*, published by LearningExpress. Lessons 17, 18, and 19 of this book pertain to these concepts.

JUST IN TIME LESSON—COORDINATE GEOMETRY

The coordinate plane is a powerful tool when used in conjunction with geometry and algebra. Study the contents of this chapter to gain important knowledge and skills to prepare for your upcoming test. Chapter 10 topics are:

- Ordered Pairs
- Midpoint
- Distance Between Two Points
- Geometric Figures in the Coordinate Plane
- Linear Equations
- Systems of Equations
- Slope—Special Relationships
- Transformations on the Coordinate Plane

ORDERED PAIRS

The coordinate plane is a grid created by the intersection of two perpendicular number lines, called the x- and y-axes. The x-axis is a horizontal number line and the *y*-axis is a vertical number line. There are four regions created by this grid and axes, as shown in the following diagram:

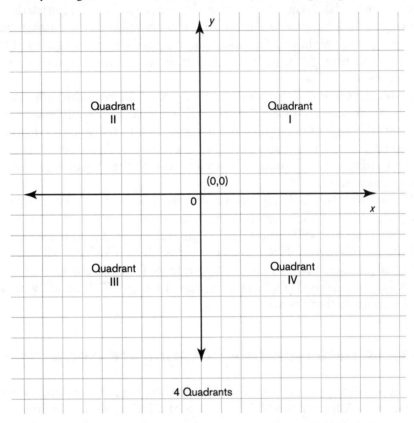

The quadrants are named by Roman numerals; Quadrant I is in the upper right corner. The other quadrants follow in a counterclockwise direction. The intersection of the axes, point *O* above, is called the *origin*. Points exist in this two dimensional space in the coordinate plane. Points are named in the coordinate plane by an ordered pair: (*x*-coordinate, *y*-coordinate). The *y*-coordinate is ALWAYS named second. The *x*-coordinate is the horizontal distance from the origin. Positive *x*-coordinates are to the right of the origin and negative *x*-coordinates are to the left of the origin. The *y*-coordinate is the vertical distance from the origin. Positive *y*-coordinates are above the origin and negative *y*-coordinates are below the origin. Each point has a unique location, as defined by its ordered pair. The coordinates for the origin are (0,0). The following graph shows some points and their ordered pair name:

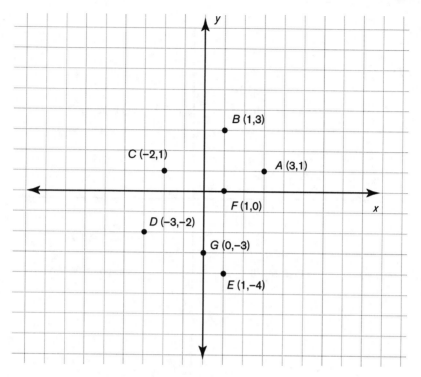

Notice the signs of the ordered pairs and where they lie in the coordinate plane:

In Quadrant I, all points have the sign (+x, +y).
In Quadrant II, all points have the sign (−x, +y).
In Quadrant III, all points have the sign (−x, −y).
In Quadrant IV, all points have the sign (+x, −y).

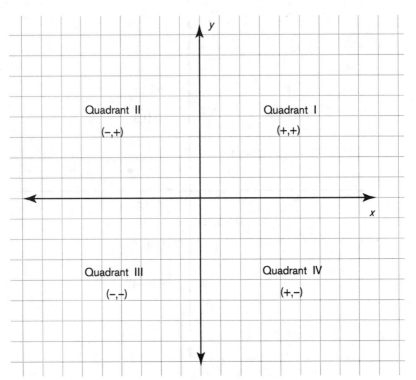

   To find the ordered pair for a point, count how far away from the origin the point is in the horizontal direction, and then count how far away from the origin the point is in the vertical direction.

Example:

What are the coordinates (ordered pairs) for points *A*, *B*, and *C* below?

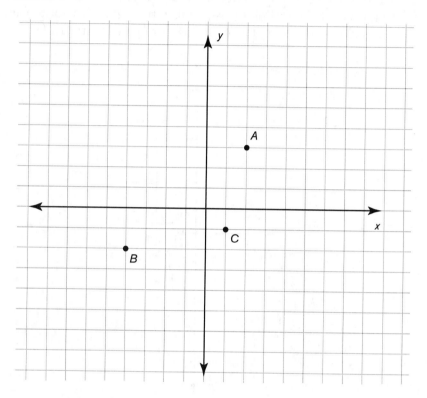

Point *A* is 2 units to the right of the origin, and 3 units above the origin. The coordinates of point *A* are (2,3). Point *B* is 4 units to the left of the origin, and 2 units below the origin. The coordinates of point *B* are (–4,–2). Point *C* is 1 unit to the right of the origin and 1 unit below the origin. The coordinates of point *C* are (1,–1).

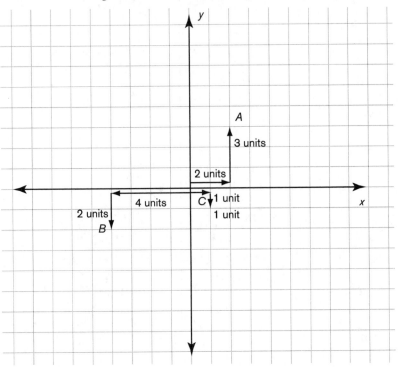

MIDPOINT

Any two points determine a segment. Every segment has a midpoint. The midpoint is exactly halfway between the two points.

RULE BOOK

The coordinates of the midpoint of a segment, given the coordinates of its endpoints as (x_1, y_1) and (x_2, y_2) is $M = (\frac{x_1 + x_2}{2}, \frac{y_1 + y_2}{2})$.

Example:
What is the midpoint of the segment with endpoints A $(-3,-5)$ and B $(-6,7)$?

Use the midpoint formula:

$M = (\frac{-3 + -6}{2}, \frac{-5 + 7}{2})$, or $M = (\frac{-9}{2}, \frac{2}{2}) = (-4.5, 1)$.
This is evident from the graph:

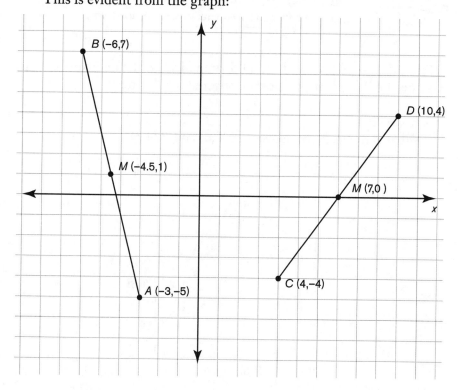

You can use the midpoint formula to find a missing coordinate.

Example:
What is the endpoint, C, of a segment whose midpoint, M, is $(7, 0)$ and other endpoint is D $(10,4)$?

Use the midpoint formula to find x_2 and y_2:

$\frac{10 + x_2}{2} = 7$	Multiply both sides by 2	$\frac{4 + y_2}{2} = 0$
$10 + x_2 = 14$	Subtract from both sides	$4 + y_2 = 0$
$x_2 = 4$		$y_2 = -4$

The other endpoint is $(4,-4)$.
This segment and midpoint are shown on the previous graph.

DISTANCE BETWEEN TWO POINTS

The distance between two points in a coordinate plane can be calculated if the coordinates of the points are known.

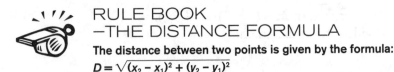

RULE BOOK
—THE DISTANCE FORMULA

The distance between two points is given by the formula:

$$D = \sqrt{(x_2 - x_1)^2 + (y_2 - y_1)^2}$$

This is an important and useful formula in coordinate geometry. It is based on the Pythagorean theorem, covered in Chapter 9.

Example:
What is the distance between A (8,6) and B (4,3)?

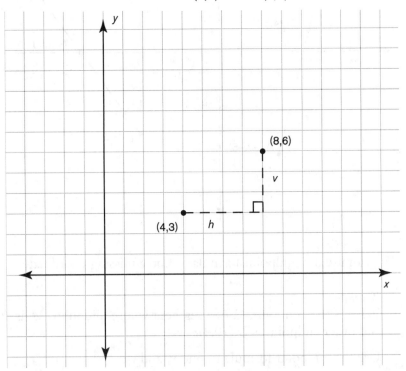

From the graph, the horizontal distance, h, and the vertical distance, v, form the legs of a right triangle. The distance between the points is the hypotenuse of the right triangle; $h^2 + v^2 = $ distance^2, and therefore $\sqrt{h^2 + v^2} = $ distance. Distance h can be found by subtracting the x-coordinate of one point from the x-coordinate of the other point. Distance v can be found the same way using the y-coordinates.

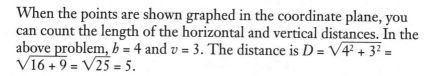
When the points are shown graphed in the coordinate plane, you can count the length of the horizontal and vertical distances. In the above problem, $h = 4$ and $v = 3$. The distance is $D = \sqrt{4^2 + 3^2} = \sqrt{16 + 9} = \sqrt{25} = 5$.

When working with the distance formula, you often must simplify radicals. To simplify a radical, factor out all perfect square factors.

Example:
Simplify $\sqrt{200}$.

$$\sqrt{200} = \sqrt{2} \times \sqrt{100} = \sqrt{2} \times 10 = 10\sqrt{2}$$

Example:
Simplify $\sqrt{252}$.

$$\sqrt{252} = \sqrt{4} \times \sqrt{63} = \sqrt{4} \times \sqrt{9} \times \sqrt{7} = 2 \times 3 \times \sqrt{7} = 6\sqrt{7}$$

Simplifying radicals is often the last step in using the distance formula.

Example:
What is the distance between $(-3, 6)$ and $(-7, -4)$?

$$D = \sqrt{(-3 - -7)^2 + (6 - -4)^2} = \sqrt{4^2 + 10^2}$$
$$D = \sqrt{16 + 100} = \sqrt{116}, \; D = \sqrt{4} \times \sqrt{29} = 2\sqrt{29}$$

SHORTCUT
If the $h(x_2 - x_1)$ and the $v(y_2 - y_1)$ are the two smaller values in a common Pythagorean triple, the distance will be the largest value in that triple set. Recall from Chapter 9, that the common Pythagorean triples are: {3, 4, 5}, {5, 12, 13}, or {8, 15, 17}, and multiples thereof.

Example:

Find the distance between points L and N in the following graph.

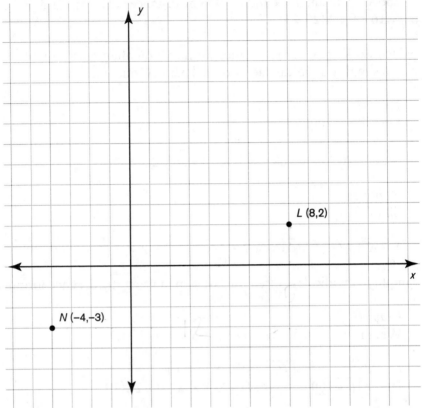

$(x_2 - x_1) = 8 - -4 = 12$ and $(y_2 - y_1) = 2 - -3 = 5$. This is part of the triple {5, 12, 13}. The distance is therefore 13 units.

GEOMETRIC FIGURES
IN THE COORDINATE PLANE

Polygons are created in the coordinate plane when various points are connected to form the sides. You may be required to determine the area, length of sides, length of diagonals, or length of altitudes of certain polygons.

Example:
What is the area of trapezoid *PQRS*?

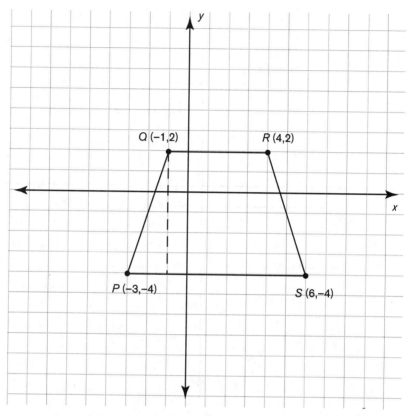

Count the squares to find the lengths of b_1 and b_2. Base b_1 is 9, $(6 - -3)$, units long; b_2 is 5, $(4 - -1)$, units long. The height can be determined by counting the perpendicular length between the bases, $2 - -4 = 6$ units long. From Chapter 6, the formula for the area of a trapezoid is: $A = \frac{1}{2}h(b_1 + b_2)$. Substitute in the values to get $A = \frac{1}{2} \times 6(9 + 5) = 42$ square units.

Example:

What is the length of the longer diagonal in parallelogram *FGHI*?

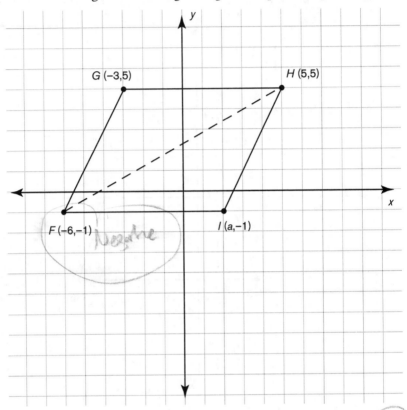

The length of the diagonal is the distance between point F $(-6,-1)$ and H $(5,5)$. Use the distance formula: $D = \sqrt{(x_2 - x_1)^2 + (y_2 - y_1)^2}$, or $D = \sqrt{(-6 - 5)^2 + (1 - 5)^2}$. This simplifies to $D = \sqrt{(-11)^2 + (-4)^2} = \sqrt{121 + 16} = \sqrt{137}$ units long.

This answer is simplified.

$\sqrt{157}$

LINEAR EQUATIONS

The coordinate plane shows the infinite solutions to an equation in two variables in picture form. For the equation $y = 2x + 3$, a table can be made to show the values that satisfy the equation:

WHEN $x =$	THEN $y =$	ORDERED PAIR SOLUTION
-2	$2(-2) + 3 = -1$	$(-2,-1)$
-1	$2(-1) + 3 = 1$	$(-1,1)$
0	$2(0) + 3 = 3$	$(0,3)$
1	$2(1) + 3 = 5$	$(1,5)$

These ordered pairs are then graphed, and joined to form a line. The line is the set of infinite points that satisfy the equation.

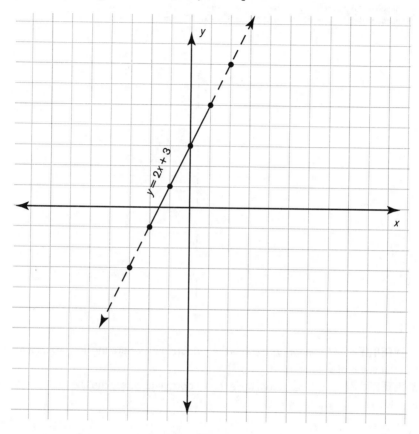

When the line is extended, as shown dotted above, other ordered pairs are defined that satisfy the equation, such as (–3,–3) and (2,7). Lines graphed in the coordinate plane have certain characteristics, such as steepness, and where the line crosses the y-axis and the x-axis.

GLOSSARY

SLOPE of a linear equation is the steepness of the line. It is the ratio of $\frac{\text{the change in the } y\text{-coordinate's value}}{\text{the change in the } x\text{-coordinate's value}}$, or $\frac{\Delta y}{\Delta x}$, $\frac{y_2 - y_1}{x_2 - x_1}$ or $\frac{\text{rise}}{\text{run}}$.

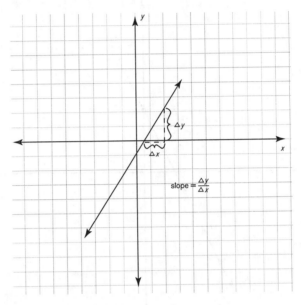

Y-INTERCEPT of a linear equation is the y-coordinate where the line crosses the y-axis. It is the value of y when $x = 0$.

X-INTERCEPT of a linear equation is the x-coordinate where the line crosses the x-axis. It is the value of x when $y = 0$.

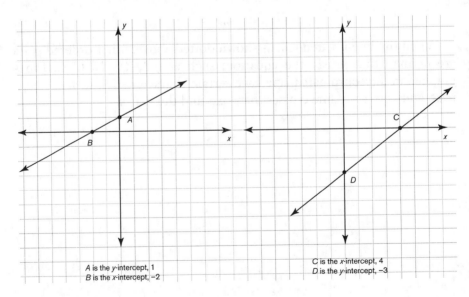

A is the y-intercept, 1
B is the x-intercept, −2

C is the x-intercept, 4
D is the y-intercept, −3

The magnitude of the slope value determines the steepness of the line. You can determine the slope of a line by first choosing two integer (integral) values for the coordinates on the line. Count the change in y, the vertical distance, and then the change in x, the horizontal distance. The slope will be $\frac{\Delta y}{\Delta x}$.

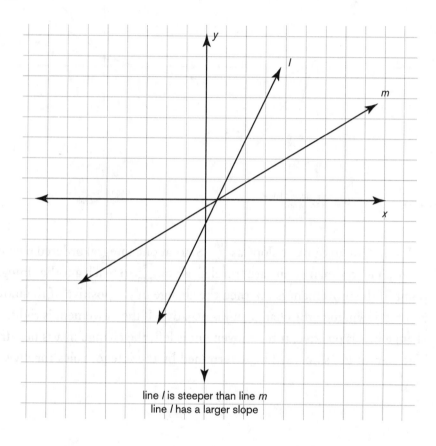

line l is steeper than line m
line l has a larger slope

The sign of the slope indicates whether the slope is "uphill" or "downhill," when moving from left to right on the graph.

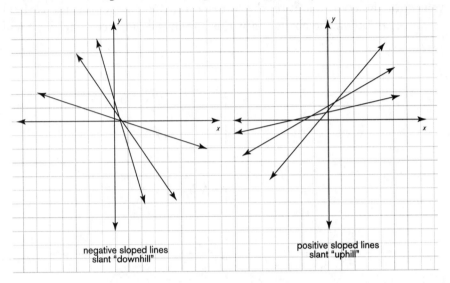

negative sloped lines
slant "downhill"

positive sloped lines
slant "uphill"

It is helpful to think of slope as $\frac{\text{rise}}{\text{run}}$. Start at one integral ordered pair and determine how many units are traveled as you $\frac{\text{rise}}{\text{run}}$ to reach another integral point. If the movement is up, then the change in y is positive; if the movement is down, then the change in y is negative. If the movement is right, the change in x is positive; if the movement is left, the change in x is negative. The sign of the slope will be determined by the integer rules for division as $\frac{\Delta y}{\Delta x}$.

Example:

What is the slope and intercepts of the following graphed line?

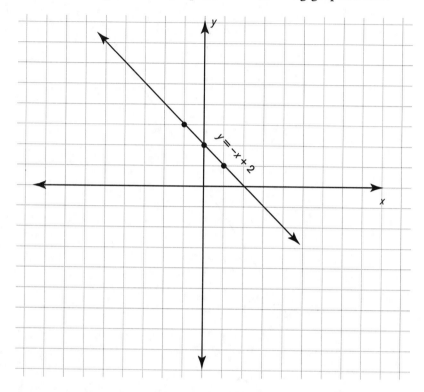

The *y*-intercept is 2, (0,2), which is the value of *y* when $x = 0$, and the *x*-intercept is 1, (2,0), which is the value of *x* when $y = 0$. Find integral points, which is often the intercepts, to determine the slope. To reach point (1,1) starting at the *y*-intercept, you rise −1 (down 1) and run +1 (right 1). So the slope is $\frac{\text{rise}}{\text{run}} = \frac{1-2}{1} - 0 = \frac{-1}{1} = -1$.

Example:
What is the slope and intercepts of the following graphed line?

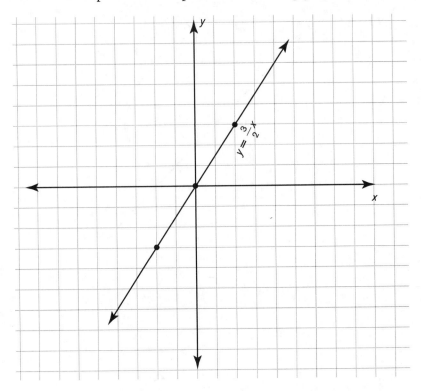

The *y*-intercept is 0, which is the value of *y* when $x = 0$, and the *x*-intercept is 0, which is the value of *x* when $y = 0$. The slope is determined from the shown integral points, from (0,0) to (2,3). Slope is $\frac{\text{rise}}{\text{run}} = \frac{+3}{+2} = \frac{3}{2}$.

The equation of a line, when in a special form called the slope-intercept form, determines the slope and *y*-intercept of the graphed equation.

GLOSSARY

SLOPE-INTERCEPT FORM of an equation is the equation solved for *y*: $y = mx + b$, where *m* and *b* are real numbers and *x* and *y* are variables.

Any equation in two variables can be expressed in slope-intercept form, by isolating the *y* variable.

Example:

Express $3x + 2y = 17$ in slope-intercept form.

Solve $3x + 2y = 17$ for *y*:

$3x + 2y - 3x = 17 - 3x$	Subtract $3x$ from both sides.
$2y = -3x + 17$	Combine like terms.
$\frac{2y}{2} = -\frac{3}{2}x + \frac{17}{2}$	Divide all terms by 2.
$y = -\frac{3}{2}x + 8.5$	

 ## RULE BOOK

When an equation is in the slope-intercept form, $y = mx + b$, *m* is the slope of the graphed equation, and *b* is the *y*-intercept of the graphed equation.

Examples:

What is the *y*-intercept of the equation $y = -7x - 5$?

The *y*-intercept is $(0,-5)$, which is -5, the *b* term. The slope of this line is -7, the *m* term.

What is the *y*-intercept of the equation $y = -7x$?

There is no *b* term, so the line crosses at $(0,0)$ and the *y*-intercept is zero.

What is the slope of the equation $x = 3$?

This equation is not in slope-intercept form; there is no *y* variable. The slope is undefined, or it is said to have no slope. This happens to be a vertical line passing through the point $(3,0)$.

What is the slope of $y = -x + 6$?

The *m* term in this example is -1, because the coefficient of the *x* term, -1, is implied when only a negative sign is shown. The slope is -1.

What is the slope of $y = 4$?

There is no *x* variable; the slope is zero. This happens to be a horizontal line passing through the point $(0,4)$.

When an equation is in slope-intercept form you can graph the solution in the coordinate plane without making a table. Use this procedure:

1. Graph the point, *b*, that is the *y*-intercept.
2. Graph a second point, using the slope, written as a fraction. The slope is $\frac{rise}{run}$. Start at the *y*-intercept and then travel up or down, and left or right as dictated by the ratio.
3. Graph another point, starting at the last point graphed, again using $\frac{rise}{run}$.
4. Connect the points into a line and label the line. Your points will form a straight line.

Example:
Graph the linear equation $y = -5x - 2$.
In this equation, $b = -2$. Graph this point as the *y*-intercept at $(0,-2)$. Use the slope of -5 and write it as a ratio, $-\frac{5}{1}$. Graph the second point using a rise of -5 (down 5) and a run of 1 (right 1) to reach the point $(1,-7)$.

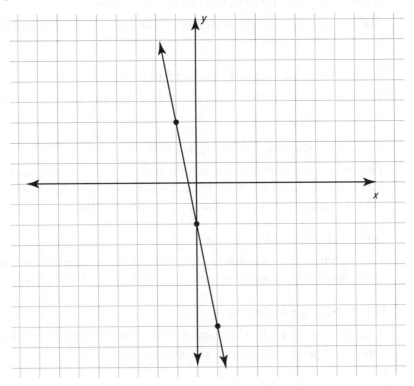

Just as you can graph a line given the equation, you can also determine the equation if given the graph:

1. Look at the graph to find the y-intercept. This is the value of b.
2. Pick two points on the graph that have integral coordinates and determine the $\frac{\text{rise}}{\text{run}} = \frac{\Delta y}{\Delta x}$ ratio. This is the value of m.
3. The equation is $y = mx + b$.

Example:
What is the equation of the following graphed line?

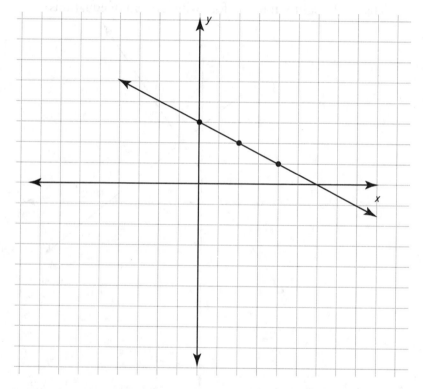

The line crosses the y-axis at positive 3, so the value of b is 3. To get from this point to another integral point on the graph (shown above), you travel down 1 (-1) and right 2 ($+2$). The slope is $\frac{\text{rise}}{\text{run}} = \frac{-1}{+2} = -\frac{1}{2}$. The equation is $y = -\frac{1}{2}x + 3$.

SYSTEMS OF EQUATIONS

A system of equations is two or more equations that have related variables. There are several ways to solve systems of equations, some of which are beyond the scope of this book. One method, solving the system graphically, is explained in this chapter. To solve a system of equations graphically, graph each equation on the coordinate plane. The solution is the intersection of the graphed lines. The solution will basically be an ordered pair to define the point of intersection.

> *Example:*
> What is the solution to the following system of equations?

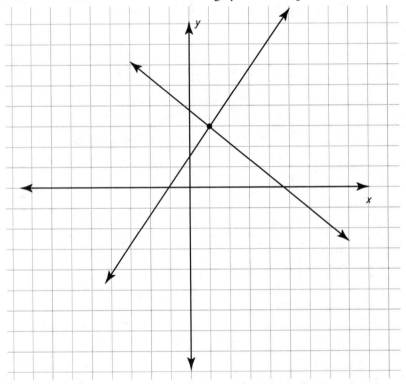

The graphed lines intersect at (1,3). This is the solution to the system. When you solve a system of equations there can be one, none, or an infinite number of solutions. When the lines intersect, they meet in one point, which is one solution. When the lines are parallel they will never meet, which means there is no solution. When the lines are equivalent their graphs fall on top of each other, which is an infinite number of solutions.

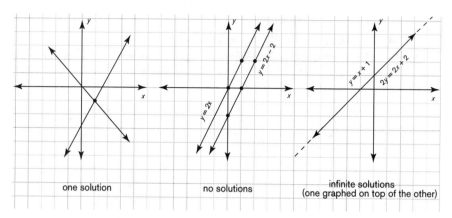

| one solution | no solutions | infinite solutions (one graphed on top of the other) |

SLOPE—SPECIAL RELATIONSHIPS

Slope is the ratio of $\frac{\text{rise}}{\text{run}}$. There are two special slopes to be considered, the slope of a horizontal line and the slope of a vertical line. A horizontal line has a slope of zero. The y-coordinate, the rise, does not change. Any fraction with zero in the numerator is equivalent to zero. A horizontal line has the form $y = b$. A vertical line has a slope that is undefined, or sometimes said to be "no slope". The x-coordinate, the run, does not change. Any fraction with zero in the denominator is undefined. A vertical line has the form $x = c$.

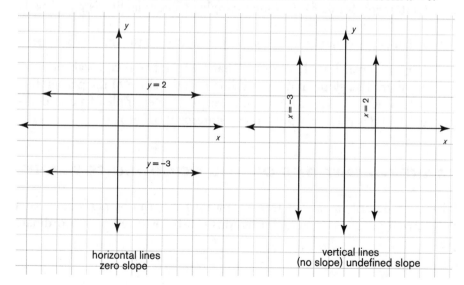

| horizontal lines zero slope | vertical lines (no slope) undefined slope |

RULE BOOK

PARALLEL LINES have equal slopes.

PERPENDICULAR LINES have slopes that are negative reciprocals of each other.

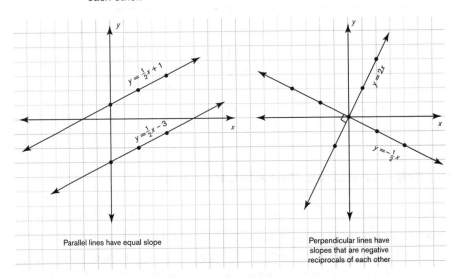

Parallel lines have equal slope

Perpendicular lines have slopes that are negative reciprocals of each other

Examples:

What is the equation of a line parallel to the line $y = -5x + 11$?

For lines to be parallel, they must have the same slope. Choose any *y*-intercept, (value for *b*). One example of a parallel line is $y = -5x + 1$.

What is the equation of a line perpendicular to the line $y = 3x - 4$?

Perpendicular lines have slopes that are negative reciprocals of one another. The slope of the given line is 3; the slope of the perpendicular line is $-\frac{1}{3}$. Choose any *y*-intercept. One example of a perpendicular line is $y = -\frac{1}{3}x$, which has a *y*-intercept of zero.

TRANSFORMATIONS ON THE COORDINATE PLANE

Chapter 8 described various transformations that can be performed on a geometric figure. These transformations can be pictured on a coordinate plane.

GLOSSARY

$T_{(A,B)}$ a translation of a graphed polygon in which each point, (x,y), is shifted to the point $(x + a, y + b)$

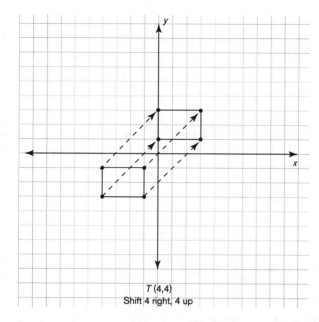

$T(4,4)$
Shift 4 right, 4 up

$\mathbf{R}_{X\text{-AXIS}}$ a reflection of a geometric figure in the x-axis. The point (x,y) becomes $(x,-y)$.

$\mathbf{R}_{Y\text{-AXIS}}$ a reflection of a geometric figure in the y-axis. The point (x,y) becomes $(-x,y)$.

$\mathbf{R}_{Y=X}$ a reflection of a geometric figure in the line $y=x$ (slope of 1, and y-intercept of 0). The point (x,y) becomes (y,x).

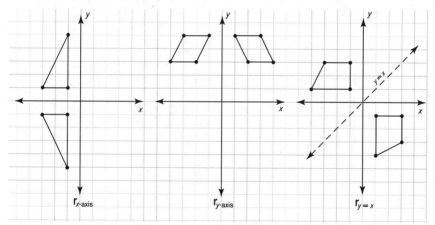

$\mathbf{R}_{P,90°}$ a rotation of a geometric figure of 90° (a quarter turn) counterclockwise around a point P. Point (x,y) becomes $(-y,x)$.

$\mathbf{R}_{P,-90°}$ a rotation of a geometric figure of 90° (a quarter turn) clockwise around a point P. Point (x,y) becomes $(y,-x)$.

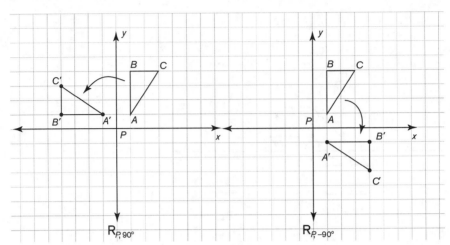

$R_{P,180°}$ a rotation of a geometric figure of 180° (a half turn) around a point P. Point (x,y) becomes $(-x,-y)$.

$R_{P,270°}$ a rotation of a geometric figure of 270° (a three-quarter turn) counterclockwise around a point P. This rotation is the same as $R_{P,-90°}$.

$R_{P,-270°}$ a rotation of a geometric figure of 270° (a three-quarter turn) clockwise around a point P. This is the same as $R_{P,90°}$.

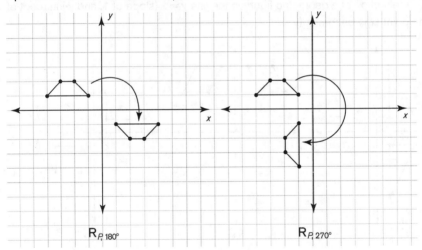

Example:
The following figure shows a translation of $\triangle ABC$ of $T_{(3,2)}$ marked $\triangle A'B'C'$ and a translation of parallelogram $EFGH$ of $T_{(-3,-1)}$ marked $E'F'G'H'$. Note that point C on the triangle, $(-1,1)$, for example, moves to point C', $(-1 + 3, 1 + 2)$, or $(2,3)$. Point E, $(3,-1)$ moves to E', $(3 - 3, -1 -1)$, or $(0,-2)$.

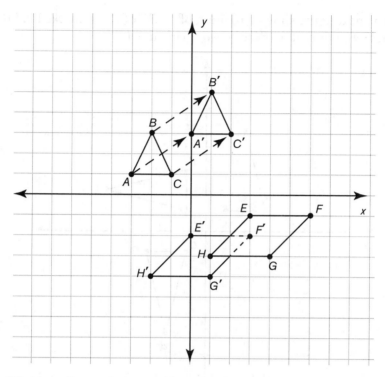

The next figure shows a reflection of $\triangle ABC$ of $r_{y\text{-axis}}$ marked $\triangle A'B'C'$ and a reflection of $r_{x\text{-axis}}$ marked $\triangle A''B''C''$.

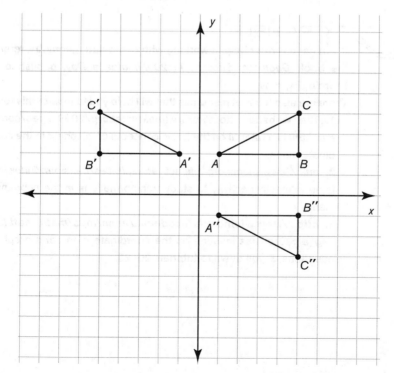

The next figure shows a rotation of $\triangle ABC$ of $R_{P,-90°}$ marked $\triangle A'B'C'$ and a rotation of $\triangle ABC$ of $R_{P,180°}$ marked $\triangle A''B''C''$.

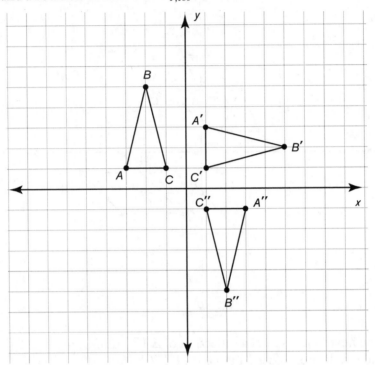

EXTRA HELP

If you need further help in working with coordinate geometry, refer to the book *Geometry Success in 20 Minutes a Day*, published by LearningExpress.

There are also Internet resources that will help you to master this topic:

1. The website *http://library.thinkquest.org/20991/geo/coord-geo.html* is a tutorial on the basic concepts of graphing in the coordinate plane.

2. *http://www.shodor.org/interactivate/activities/transform/index.html* is an interactive activity to study transformations in the coordinate plane.

3. *http://www.wtamu.edu/academic/anns/mps/math/mathlab/beg_algebra/* has lessons on the coordinate plane and graphing equations. Scroll down to tutorials 20–25.

TIPS AND STRATEGIES

- The coordinate plane is a grid formed by horizontal and vertical number lines that intersect at the point called the origin.
- The coordinate plane is divided into four quadrants.
- An ordered pair (x,y) defines a point in the coordinate plane as the horizontal distance, x, from the origin and the vertical distance, y, from the origin.
- In Quadrant I the coordinates have the sign (+,+).
- In Quadrant II the coordinates have the sign (–,+).
- In Quadrant III the coordinates have the sign (–, –).
- In Quadrant IV the coordinates have the sign (+,–).
- The midpoint is the point halfway between two endpoints of a segment.
- The distance between two points in the coordinate plane is found by using the distance formula.
- Geometric polygons reside in the coordinate plane.
- Area, side length, diagonal length, and height can be calculated using coordinates of endpoints.
- The slope of a linear equation is the ratio of $\frac{\Delta y}{\Delta x} = \frac{\text{rise}}{\text{run}}$.
- The y-intercept is the value of y when $x = 0$, where a graphed equation crosses the y-axis.
- The slope-intercept form of a line, $y = mx + b$, determines the slope and the y-intercept of a graphed line.
- A system of equations can be solved graphically by finding the intersection of the graphed lines.
- Parallel lines have equal slopes.
- Perpendicular lines have slopes that are negative reciprocals of each other.
- Horizontal lines have a slope of zero.
- Vertical lines have no slope, or a slope that is undefined.
- Geometric figures can be transformed on the coordinate plane.

PRACTICE QUIZ

1. In which quadrant does the point (–4,–6) lie?
   a. Quadrant I
   b. Quadrant II
   c. Quadrant III
   d. Quadrant IV
   e. Quadrant V

2. What are the coordinates of the point *M* shown below?

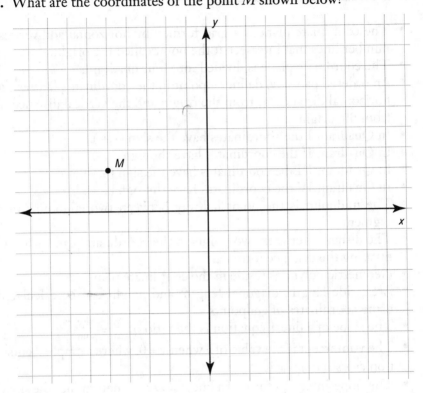

   a. (–5,2)
   b. (5,2)
   c. (2,–5)
   d. (–2,5)
   e. (–2,–5)

3. What are the coordinates of the midpoint of the segment joined by (–7,–4) and (–1,5)?
   a. (–4,4.5)
   b. (4.5,–4)
   c. (–3,0.5)
   d. (–3,4.5)
   e. (–4,0.5)

4. One endpoint of a segment is (–8,–2). What is the other endpoint, if the midpoint of the segment is (0,0)?
   a. (2,8)
   b. (8,2)
   c. (–4,–1)
   d. $(\frac{1}{8},\frac{1}{2})$
   e. (4,1)

5. What is the distance between the points (1,4) and (4,8)?

   a. $\sqrt{5}$

   b. 13

   c. $\sqrt{13}$

   d. 5

   e. 17

6. What is the length of the diagonal in the graph of the square below?

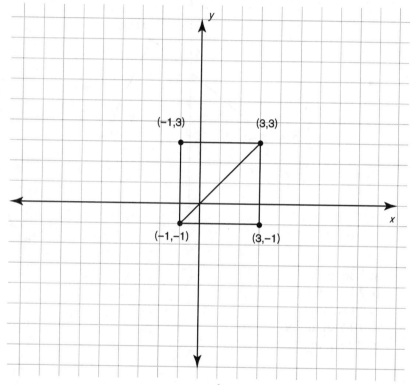

   a. $4\sqrt{2}$

   b. 8

   c. $\sqrt{8}$

   d. 32

   e. 4

7. In the following graph, what is the area of $\triangle PQR$?

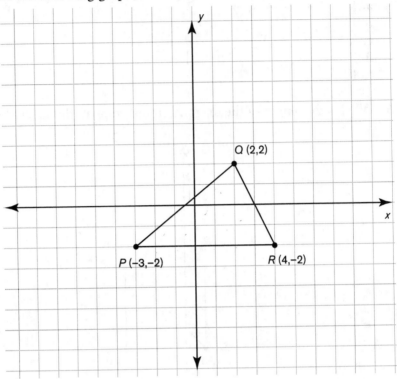

a. 7 sq. units
b. 28 sq. units
c. 14 sq. units
d. 10.5 sq. units
e. 0.5 sq. units

8. What is the length of the height of trapezoid *ABCD* on the following graph?

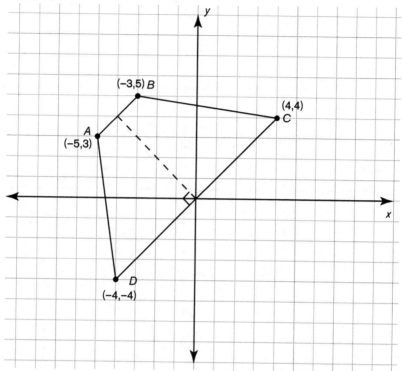

a. $\sqrt{37}$ units
b. 6 units
c. $4\sqrt{2}$ units
d. 32 units
e. $\sqrt{8}$ units

9. What is the length of the segment whose endpoints are (–2,–3) and (6,–1)?
a. $2\sqrt{17}$ units
b. 68 units
c. $\sqrt{32}$ units
d. 32 units
e. $4\sqrt{5}$ units

10. What is the equation of the line shown in the following graph?

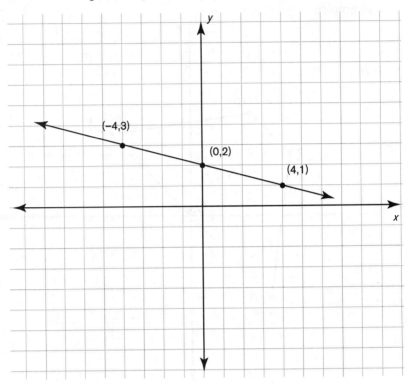

a. $y = -2x + 2$
b. $y = -\frac{1}{4}x + 2$
c. $y = 2x - \frac{1}{4}$
d. $y = -4x - 2$
e. $y = \frac{1}{2}x + 2$

11. What is the equation of a line that is parallel to the graph of line
$y = 3x - 6$?
  a. $y = -6x + 3$
  b. $y = -\frac{1}{3}x + 2$
  c. $y = 3x + 6$
  d. $y = -3x - 6$
  e. $y = \frac{1}{3}x - 6$

12. Which is the slope of the following graphed line?

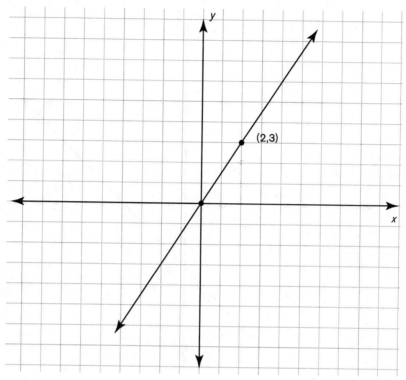

a. $\frac{2}{3}$

b. 5

c. 3

d. $\frac{3}{2}$

e. −5

13. What is the *y*-intercept of the following graphed line?

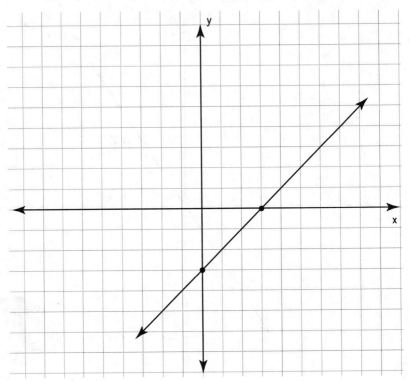

a. $\frac{2}{3}$

b. $\frac{3}{2}$

c. 3

d. 2

e. –3

14. What is the equation of a line perpendicular to the graphed line in the following graph?

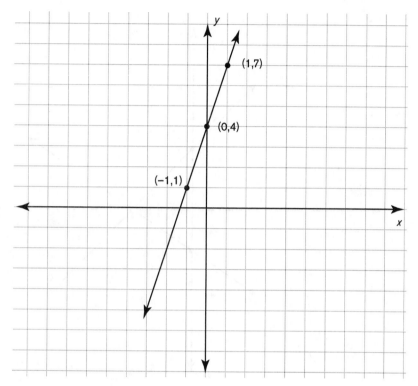

a. $y = -3x + 2$
b. $y = -\frac{1}{3}x + 2$
c. $y = 3x - 2$
d. $y = x - 2$
e. $y = -x + 4$

15. What is the solution to the system of equations shown in the following graph?

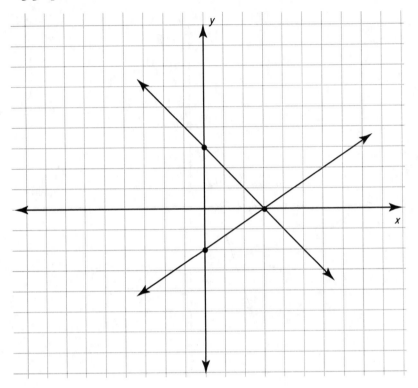

a. (3,0)
b. (0,3)
c. (−2,0)
d. (0,−2)
e. (0,−5)

16. What is the slope of the following line?

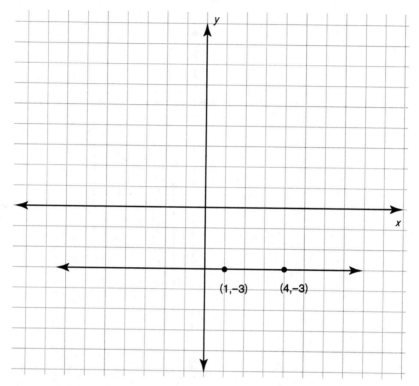

(1,−3)   (4,−3)

a. 0
b. undefined
c. 2
d. $\frac{3}{4}$
e. 1

17. What is the area of the parallelogram shown below?

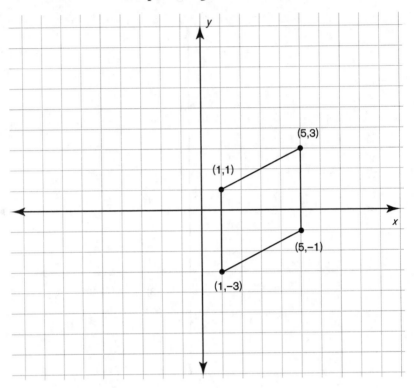

a. $2\sqrt{5}$ square units
b. 16 square units
c. 20 square units
d. 8 square units
e. 12 square units

18. What is the length of the diagonal of rectangle *ABCD* shown following?

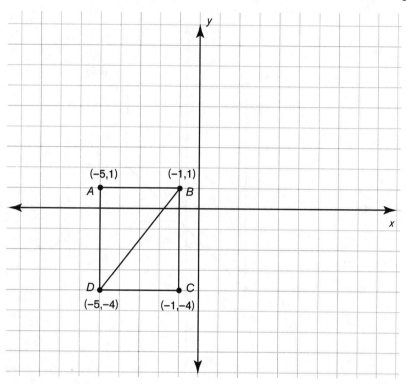

a. $\sqrt{61}$ units
b. 20.5 units
c. 30.5 units
d. $\sqrt{41}$ units
e. $\sqrt{50}$ units

19. What is the area of trapezoid *LMNO*?

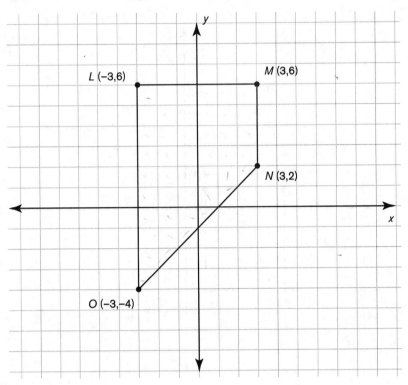

a. 40 square units
b. 48 square units
c. 54 square units
d. 24 square units
e. 42 square units

20. Which transformation is shown in the following figure from *ABCD* to *A'B'C'D'*?

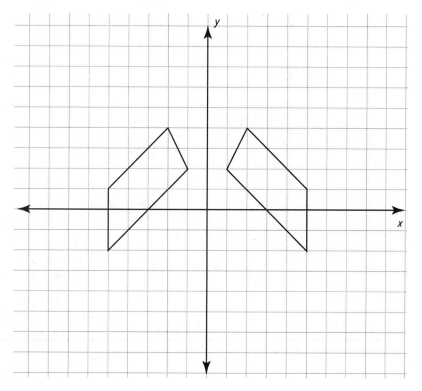

a. $r_{y = x}$
b. $R_{P,90°}$
c. $r_{x\text{-axis}}$
d. $r_{y\text{-axis}}$
e. $T_{(10,0)}$

21. Which of the following figures shows a rotation of 90° clockwise, from *ABC* to *A′B′C′*?

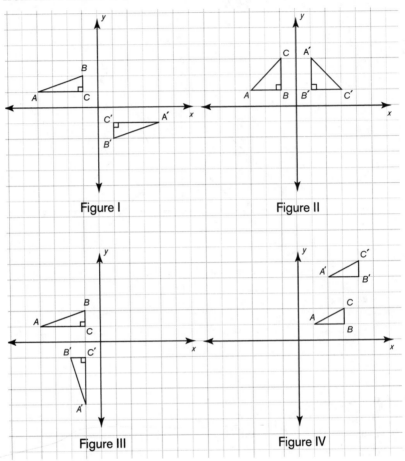

Figure I

Figure II

Figure III

Figure IV

a. Figure I
b. Figure II
c. Figure III
d. Figure IV
e. none of the above

22. What is the equation of the following graphed line?

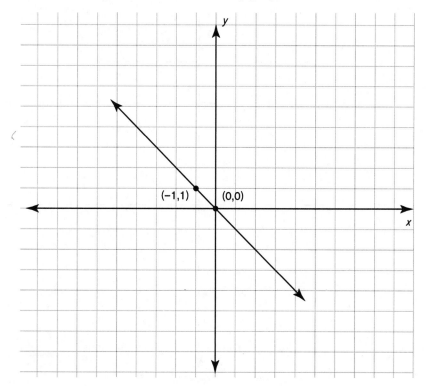

a. $y = x$
b. $y = -x + 1$
c. $y = -x - 1$
d. $y = -x$
e. The equation is not shown.

23. What is the equation of the following graphed line?

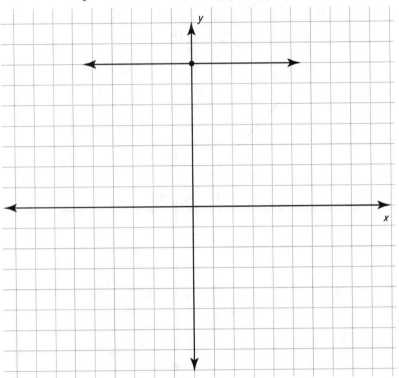

a. $y = 7$
b. $y = -7$
c. $x = -7$
d. $x = 7$
e. $y = x + 7$

24. What is the transformation shown in the following graph of *ABCD* and its image *A'B'C'D'*?

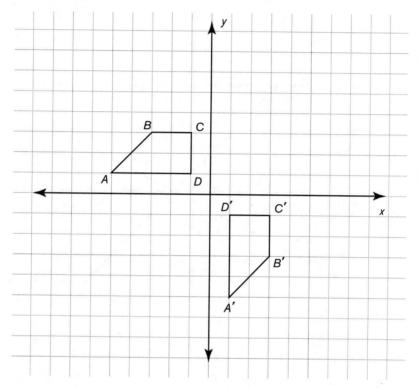

a. $R_{P,-180°}$
b. $T_{(-2,3)}$
c. $T_{(3,-2)}$
d. $r_{x\text{-axis}}$
e. $r_{y=x}$

25. What is the solution to the following graphed system of equations?

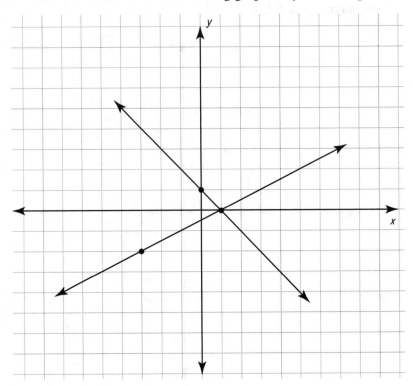

    a. (1,0)
    b. (0,1)
    c. (−3,−2)
    d. (3,−2)
    e. (−2,−3)

ANSWERS

1. c. This point has coordinates that are both negative, and thus lies in Quadrant III.

2. a. Point M is five units to the left of the origin, so the x-coordinate is −5. Point M is two units above the origin, so the y-coordinate is 2. The coordinates are (−5,2).

3. e. Use the midpoint formula: $M = (\frac{x_1 + x_2}{2}, \frac{y_1 + y_2}{2})$:
$M = (\frac{-7 + -1}{2}, \frac{-4 + 5}{2}) = (\frac{-8}{2}, \frac{1}{2})$. This is (−4,0.5).

4. b. Use the midpoint formula: $M = (\frac{x_1 + x_2}{2}, \frac{y_1 + y_2}{2})$ to solve for the variables x_2 and y_2:

$\frac{-8 + x_2}{2} = 0$	Use the midpoint formula.	$\frac{-2 + y_2}{2} = 0$
$-8 + x_2 = 0$	Multiply both sides by 2.	$-2 + y_2 = 0$
$x_2 = 0 + 8$	Isolate the x and y.	$y_2 = 0 + 2$
$x_2 = 8$	Combine like terms.	$y_2 = 2$

The coordinates are (8,2).

5. d. Use the distance formula: $D = \sqrt{(x_2 - x_1)^2 + (y_2 - y_1)^2}$

$D = \sqrt{(1 - 4)^2 + (4 - 8)^2}$, or $D = \sqrt{(-3)^2 + (-4)^2}$, or $D = \sqrt{9 + 16} = \sqrt{25} = 5$ units.

6. a. Use the coordinates of opposite vertices and the distance formula to find the length of the diagonal: $D = \sqrt{(x_2 - x_1)^2 + (y_2 - y_1)^2}$, and the vertices are (3,3) and (–1,–1).

$D = \sqrt{(3 - -1)^2 + (3 - -1)^2}$, or $D = \sqrt{(4)^2 + (4)^2}$, or $D = \sqrt{16 + 16} = \sqrt{32}$. This can be simplified to $4\sqrt{2}$ units.

7. c. Use the formula for the area of a triangle: $A = \frac{1}{2}bh$. Count the units for b (the base) and h (the height). The base is $4 - -3 = 7$ units long. The height, h, is $2 - -2 = 4$ units long. Substitute in these values to get: $A = \frac{1}{2} \times 7 \times 4$, or 14 square units.

8. c. Find the length of the height of the trapezoid by using the coordinates of the endpoints of the height, which are (0,0) and (–4,4). The distance formula is: $D = \sqrt{(x_2 - x_1)^2 + (y_2 - y_1)^2}$, or $D = \sqrt{(0 - -4)^2 + (0 - 4)^2}$, or $D = \sqrt{(4)^2 + (-4)^2}$, or $D = \sqrt{16 + 16} = \sqrt{32}$. This simplifies to $4\sqrt{2}$ units long.

9. a. Use the distance formula: $D = \sqrt{(x_2 - x_1)^2 + (y_2 - y_1)^2}$, or $D = \sqrt{(-2 - 6)^2 + (-3 - -1)^2}$, or $D = \sqrt{(-8)^2 + (-2)^2} = \sqrt{68}$. This simplifies to $2\sqrt{17}$ units long.

10. b. The graphed line crosses the y-axis at (0,2), so the y-intercept is 2. The slope can be calculated from the points (0,2) and (4,1), using $\frac{rise}{run} = \frac{y_2 - y_1}{x_2 - x_1} = \frac{2 - 1}{0 - 4} = \frac{1}{-4} = -\frac{1}{4}$. The equation is $y = -\frac{1}{4}x + 2$.

11. c. The equation of a line parallel to the given equation will have the same slope. The only equation that has the same slope, which is 3, is choice **c.** When an equation is in the form $y = mx + b$, such as these, the slope is the coefficient of the x variable.

12. d. The slope can be calculated from the points (0,0) and (2,3), using $\frac{\text{rise}}{\text{run}} = \frac{y_2 - y_1}{x_2 - x_1} = \frac{3 - 0}{2 - 0} = \frac{3}{2}$.

13. e. The y-intercept is the value of y when $x = 0$, or where the graphed line crosses the y-axis. This is at the point (0, –3). The y-intercept is –3.

14. b. The slope of a perpendicular line will have a slope that is the negative reciprocal of the given equation. The slope of the given equation is calculated by using two of the integral points shown, such as (0,4) and (–1,1). Calculate the slope: $\frac{\text{rise}}{\text{run}} = \frac{y_2 - y_1}{x_2 - x_1} = \frac{1 - 4}{-1 - 0} = \frac{-3}{-1}$, which is 3 . The negative reciprocal is $-\frac{1}{3}$. Choice **b** is the only choice with this slope, the coefficient before the variable x when the equation is in the form $y = mx + b$.

15. a. The solution to the system is the coordinates of the point of intersection. This point is three units to the right of the origin, so the x-coordinate is 3, and zero units from the origin in the vertical direction, so the y-coordinate is 0. The coordinates are (3,0).

16. a. The slope of a horizontal line is always zero. There is a zero change in the y-coordinates, which is the numerator of the slope ratio.

17. b. Use the formula for the area of a parallelogram: $A = bh$. Use the vertical side as the base, and count the units in length (using the y-coordinates). It is $1 - -3 = 4$ units long. The height is the horizontal distance between the points (using the x-coordinates). It is $5 - 1 = 4$ units high. The area is $4 \times 4 = 16$ square units.

18. d. Use the distance formula on the coordinates of the opposite vertices, such as (–1, 1) and (–5, –4). The distance formula is $D = \sqrt{(x_2 - x_1)^2 + (y_2 - y_1)^2}$, or $D = \sqrt{(-5 - -1)^2 + (-4 - 1)^2}$, or $D = \sqrt{(-4)^2 + (-5)^2} = \sqrt{16 + 25} = \sqrt{41}$ units long.

19. e. Use the formula for the area of a trapezoid: $A = \frac{1}{2}h(b_1 + b_2)$. The bases are vertical in this trapezoid. Count the units for b_1 (the base), b_2 (the other base) and the h (the height). Base 1, b_1, (using y-coordinates) is $6 - -4 = 10$ units long. Base 2, b_2, (using y-coordinates) is $6 - 2 = 4$ units long. The height, h, (using the x-coordinates) is the top side of the trapezoid, $3 - -3 = 6$ units long. Substitute in these values to get: $A = \frac{1}{2} \times 6(10 + 4)$, or $A = \frac{1}{2} \times 84 = 42$ square units.

20. d. This is a reflection, or a flip, over the y-axis. This is denoted by $r_{y\text{-axis}}$.

21. b. Figure II shows a rotation of $90°$ clockwise. Figure I is a rotation of $180°$. Figure III is a rotation of $-270°$, and Figure IV is a translation of 1 unit right and 3 units down.

22. d. The graphed line crosses through the origin, so the y-coordinate is 0. The slope can be calculated from the points $(0,0)$ and $(-1,1)$, using $\frac{\text{rise}}{\text{run}} = \frac{y_2 - y_1}{x_2 - x_1} = \frac{1 - 0}{-1 - 0} = \frac{1}{-1} = -1$. The equation, in $y = mx + b$ form, is $y = -x$, since the y-intercept is 0 and the slope, -1 is implied by writing $-x$.

23. a. This is the graph of a horizontal line, which has the form $y = b$. The y-intercept is 7, so the equation is $y = 7$.

24. e. This is the reflection of the polygon over the line $y = x$, denoted by $r_{y = x}$. The line of reflection is shown:

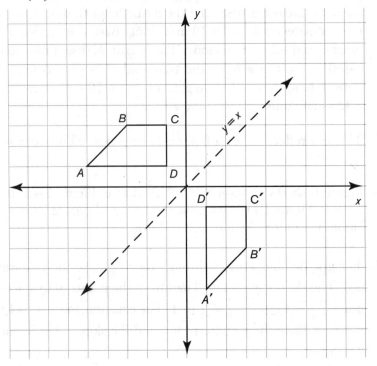

25. a. The solution to the system is the coordinates of the point of intersection. This point is one unit to the right in the horizontal direction, so the x-coordinate is 1, and zero units above the origin, so the y-coordinate is 0. The coordinates of the solution point are (1,0).